Valve and Actuator Technology

Valve and Actuator Technology

Wayne Ulanski

McGraw-Hill, Inc.
New York St. Louis San Francisco Auckland Bogotá
Caracas Hamburg Lisbon London Madrid
Mexico Milan Montreal New Delhi Paris
San Juan São Paulo Singapore
Sydney Tokyo Toronto

Library of Congress Cataloging-in-Publication Data

Ulanski, Wayne.
Valve and actuator technology / Wayne Ulanski.
p. cm.
Includes index.
ISBN 0-07-019477-7
1. Valves. 2. Actuators. I. Title.
TS277.U34 1991
621.8'4—dc20 90-21090
CIP

1 2 3 4 5 6 7 8 9 0 DOC/DOC 9 5 4 3 2 1

ISBN 0-07-019477-7

The sponsoring editor for this book was Robert W. Hauserman, the editing supervisor was Kimberly A. Goff, and the production supervisor was Suzanne W. Babeuf.

This book was set in Century Schoolbook. It was composed by McGraw-Hill's Professional Publishing composition unit.

Printed and bound by R. R. Donnelley and Sons Company.

Cover was designed by J. Madison Advertising of Hackensack, New Jersey.

This book is dedicated to the society of process control engineers worldwide and to Robin, Rebecca, and Emily.

Contents

Preface

"The practical genius is not someone who knows everything but knows where to find it."

The process industry, which represents nearly 13 percent of the U.S. gross national product, includes chemical and petrochemical manufacturing, oil refining, pulp and paper, pharmaceuticals, foods, and many other industries that directly or indirectly provide energy, medicines, nutrition, and other goods that enhance the quality of life. These industries have one thing in common, they all process liquids and they all use valves and actuators to do so.

Valves and actuators are used to permit or restrict the flow of a fluid, modify its rate, blend or distribute a fluid, isolate equipment, fill tanks, or serve as safety devices in the event of a fire or other emergency.

Valves have proven to be one of the most important and prevalent elements in today's flow control scheme. Unfortunately, simply because of the lack of knowledge about valves, actuators, and other automation technology, many industrial manufacturers are continuing to use outdated, less-efficient flow control techniques. With the flood of new technology inundating our industry, the amount of new information and the range of new products have become so diverse that narrowing the field and selecting the best product for the particular application have become costly and time-consuming tasks.

The challenge for today's process engineer is to gain the knowledge and expertise to arrange the most effective process control system by defining specific needs, selecting and matching the appropriate components, and building the system with consideration of proper design and performance compatibility.

Until *Valve and Actuator Technology*, the only sources of information on valves, actuators, and their designs and use were user's manuals and product literature, published by manufacturers about their equipment; and technical articles written by engineers at a highly technical level. This book—the first of its kind dedicated exclusively to valves and actuators—supplements these other sources by integrating information necessary for designing, selecting, and interfacing valves and automated valves with today's process control requirements.

In writing this book, I have assembled a wide variety of topics. By design, it includes information often taken for granted, for less technically skilled readers, as well as advanced design, performance, and application methods, for those who are more familiar with the role of valves and actuators in process control. In addition, the range of information covered makes the book especially well suited for use by college, universities, and technical schools. It is my opinion that readers at all levels will find this book to be a valuable reference tool and useful design selection guide.

Because each chapter of the book builds upon information presented in every preceding chapter, it is important that new readers comprehend early material before advancing to later chapters. A glossary and index are provided to facilitate the usefulness of this book and to aid in understanding unfamiliar terminology and concepts.

I hope that the knowledge gained from this book will enable readers to perform their jobs more effectively, to understand the manufacturer's literature better, to become more aware of the potential benefits of the current technology, and to make more informed and creative decisions in valves and actuators and their applications.

Wayne Ulanski

Acknowledgments

In preparing this work, I have used many resources including manufacturers literature, articles, technical papers, reference books, and other published materials. The following is a list of the companies and people I wish to acknowledge for their help and the use of their materials.

Automax Incorporated

Colonial Engineering Incorporated

Conameter Corporation

Crane Company

Demco Division of Cooper Industries

DeZurik

Dresser/Nilcor

The Duriron Company

Dynamic Air/Posi-Flate

EIM Controls

El-o-matic Incorporated

Fisher Controls International, Incorporated

Fujikin of America, Incorporated

Garlock, Incorporated

G. H. Bettis, Incorporated

Instrument Society of America

ITT Engineered Valves

Kerotest Manufacturing Corporation

Keystone International

KTM, Incorporated

Larad Equipment

Lunkenheimer Canada

Manufacturers Standardization Society

MCF Valves

MTL Limited

Neles-Jamesbury Incorporated

Orbit Valve Company

Pittsburgh Brass Manufacturing Company

Posacon, Incorporated

Posi-Seal International, Incorporated

Quality Controls, Incorporated

Randon Research

Red Valve Company, Incorporated

Rockford Controls, Incorporated

Rotork Controls, Incorporated

R. Stahl, Incorporated

Skil Controls

Stockham, Incorporated

SVF, Incorporated

Target Rock Corporation

Valve Manufacturers Association of America

Worcester Controls

Xomox Corporation

Control-Valve Selection and Sizing by Les Driskell (I.S.A. 1983)

Control Valve Handbook Fisher Controls International (1977)

I'd also like to thank the staff at McGraw-Hill Professsional Publishing: Bob Hauserman, Kim Goff, Rachel Handler, and everyone else there who contributed directly or indirectly to this project. I must also mention Claudia Dunlop whose asssistance on the manuscript was especially valuable. Finally, a special thanks to Stephanie Finn who helped make it possible.

Wayne Ulanski

Valve and Actuator Technology

Part

1

Valves

Chapter

1

Valve Selection Basics

1.1 General Introduction to Valves

The valve existed in the Egyptian, Greek, and Roman cultures centuries ago, but the modern history of the valve industry parallels the industrial revolution, when in 1705 the first industrial steam engine was invented. Because steam built up pressures that had to be contained and regulated, valves suddenly acquired a new importance. As other new inventions came along, valves became more sophisticated, more reliable, and more widely used in numerous industrial sectors.

Today, industry's requirements continue to force an evolution in valve design. Higher pressure, in a wider range of temperatures; exotic, hazardous, and expensive fluids; automation; response; and simplicity are all contributing to tomorrow's valve designs.

The valves most commonly used throughout industry may be categorized as linear (multiturn) or rotary (part-turn) and are listed in the following table.

Linear valves	Rotary valves
Gate	Ball
Globe	Plug
Diaphragm	General-purpose butterfly
Pinch	High-performance butterfly

There are also many unique valve designs intended for unique services or that combine the features of other valves for improved performance. These hybrid valves are discussed later in this book and include

Multiturn thrust-seated ball valve

Quarter-turn pinch valve

Torque-seated all-metal butterfly valve

Solenoid-actuated globe valve

Inflatable resilient seated butterfly valves

What is a valve? A valve is a pressure-containing mechanical device used to shut off or otherwise modify the flow of a fluid that passes through it. The action of a valve is caused by moving the closure element (ball, plug, gate, disc, etc.) which is attached to a stem located outside the body. This very basic definition of a valve is intended to reveal that indeed a valve is a very simple device. It has a body to contain the fluid pressure, a closure element to shut off or modify the flow through the body, and finally a stem to change the control position of the closure element.

In today's industrial environment valves are used to satisfy a variety of process and production requirements. Following is a brief discussion of what I call the five basic valve functions.

1. *On-off service.* (See Fig. 1.1.) This is basically a flow/no flow service. Valves in this function are used to block the flow of the process media. They are sometimes called a block valve. This is by far the most common use for valves.

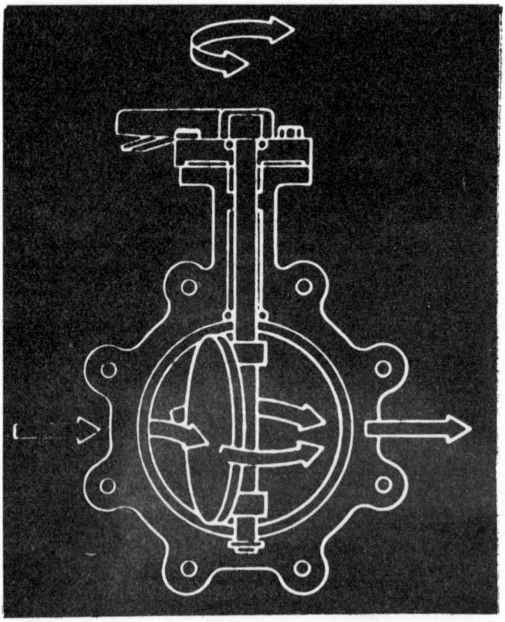

Figure 1.1 Valves in on-off service provide full flow and incomplete shutoff in a variety of applications.

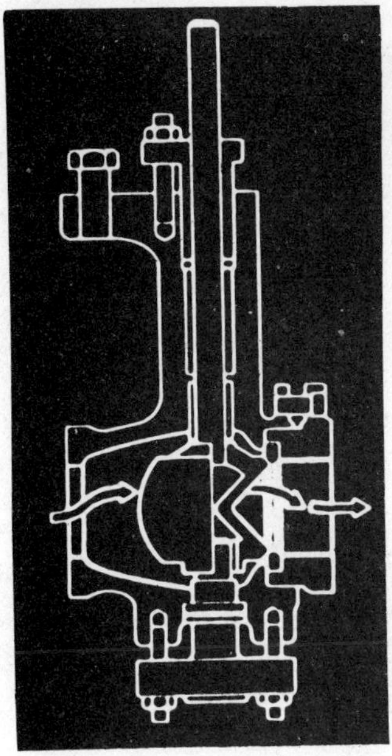

Figure 1.2 When valves are used to modify flow or throttle, they are referred to as control valves.

2. *Flow Control.* (See Fig. 1.2.) Valves in this service are used to modify the flow of a fluid by changing the flow rate or creating a pressure drop to meet the process requirements. As there is an abundance of information available on traditional cage-guided globe and other linear control valves this book will deal with the role rotary valves play in the flow control function.

3. *Divert flow.* (See Fig. 1.3.) Just as the term implies, the flow of a fluid may be diverted or distributed with the use of multiported valves. Three-way valves are the most common type of valve for this service, but four- and five-way valves are also available, and some may be manifolded to further expand the porting arrangements and flow patterns.

4. *Two-stage shutoff.* (See Fig. 1.4.) Batch control systems require two-stage control. This may be achieved with the use of two valves arranged in a parallel bypass system. When both valves are open full, flow is provided. If one is shut off, a reduced flow (dribble) condition exists to slowly reach the required batch level. No flow occurs when both valves are closed. There are some control packages that can achieve this with a single valve.

5. *Blending.* (See Fig. 1.5.) When two valves are located on a common tee and are operated concurrently by a single actuator, they are often used to blend two fluids. A common application is for tempera-

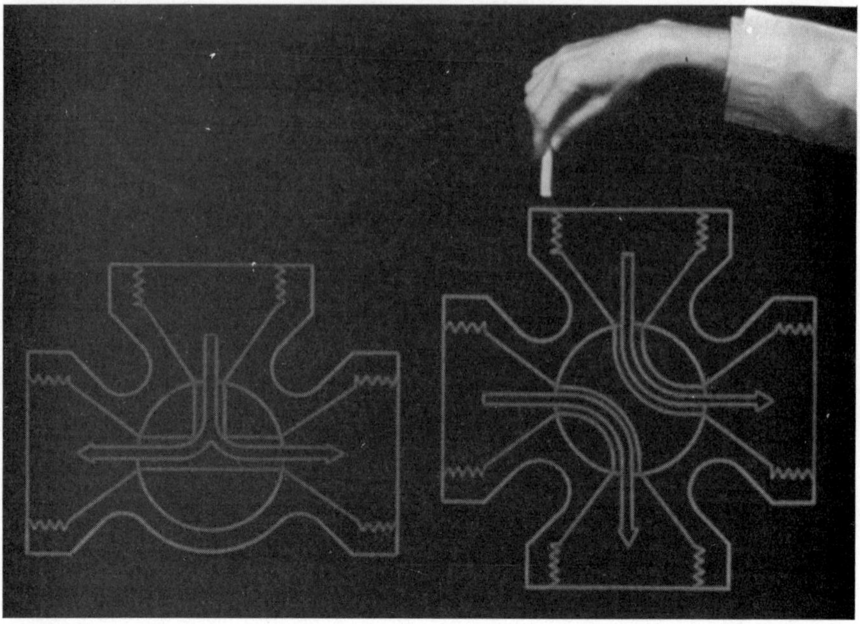

Figure 1.3 Multiported valves are used to divert, blend, switch, and isolate fluids.

ture control on a heating and cooling system in commercial buildings where hot and cold water are blended. There are also some process control systems where the same function may be required. Tee assemblies are ideal when diverting flow is required in larger piping systems.

The valve user may select from many manufacturers, materials, and designs to meet flow and process requirements. What complicates things for valve manufacturers and users, however, is not the function of a valve but under what conditions it must perform. The controlled medium may be a gas, a highly corrosive liquid, toxic or radioactive material, or any of a number of other fluids. Valves must withstand pressures from vacuum to several thousand pounds per square inch as well as cryogenic to very high temperatures. Some valves must cycle in just a few seconds repeatedly over long periods of time, while others may remain in one position for 90 percent of their life. Depending on the conditions, valve life may be a few hours or many years, and maintenance may be required after a single cycle or not required for many thousands of cycles. Applications vary widely, as does valve performance.

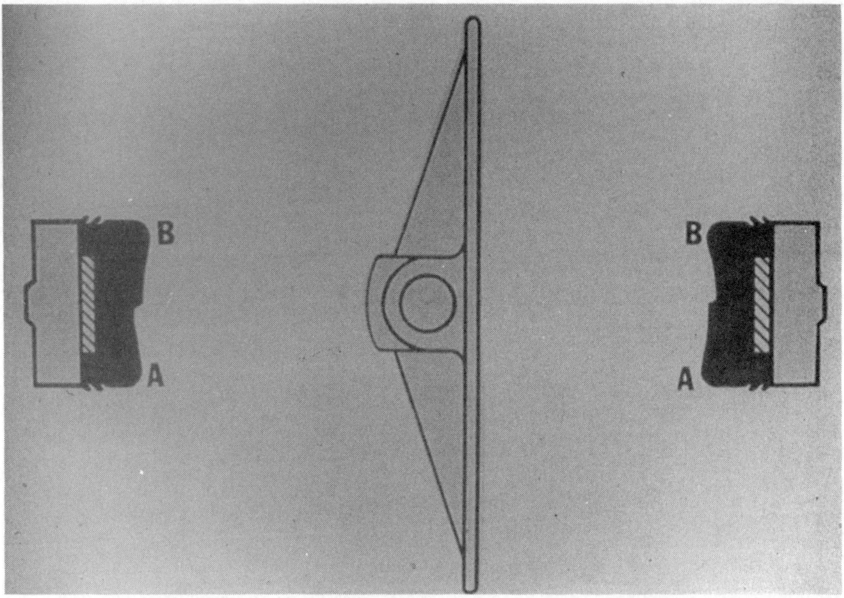

(a)

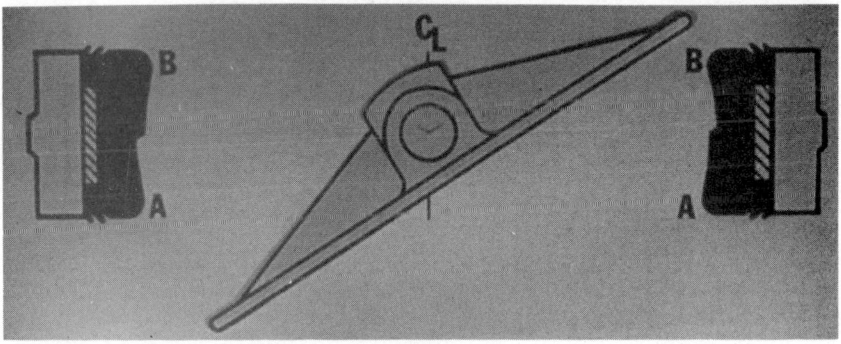

(b)

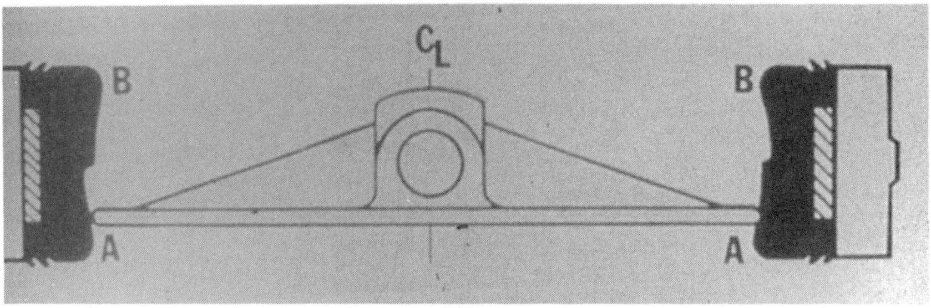

(c)

Figure 1.4 In some filling operations, valves are used to provide two-stage shutoff for more accurate control.

Figure 1.5 Tee assembly includes two valves, driven by a single actuator.

Valve selection basics

When making a valve selection, here are some important questions to ask.

Does the valve meet the specifications required for your process?

What type of valve best suits your needs: gate, globe, ball, butterfly, check, pinch, etc?

How many worker-hours does the valve take to install?

Are your employees skilled enough to install the valve or is outside help required?

What will outside help cost?

Is there a dealer or distributor near you who you can contact for parts?

How long will you have to wait for parts?

Can you rely on the distributor for service?

If the wait for parts means your process is down, how much will it cost in lost product and time?

Beware of low cost; it may mean cheap construction. Make sure the valve is sturdy and the stem is blowout proof. Will the valve be automated? Do you need a fail-safe capability in case of electrical or system failure?

Progress in design has put a great variety of valve and actuator types, each with some special qualification for service, at the engineer's fingertips. From these the engineer may choose the right automated valve package to provide dependable and economical performance.

Because the valve is in contact with the fluid that is either the final or intermediate product of a process, it is often the most carefully selected element of an automated valve package. But before an actuator selection is made, the valve type, material, etc., must be determined, as this may affect the amount of torque required to automate a valve. Having done this the actuator style (rotary or linear), torque (or thrust), output, and type (manual, electric, pneumatic, hydraulic, etc.) may be addressed.

Certain essential information must be known and analyzed before one can select the best valve for the job. Some important questions to ask are

What is the function of the valve? On-off only? Control? Fail-safe? Frequent or infrequent operation?

How will it be operated: manually, pneumatically, or electrically?

What is the process medium? Is it corrosive or abrasive? What is its volume and velocity?

Is the valve being considered available in all materials?

What about pressure and temperature? Depending upon the type of material used for a given media the pressure and temperature rating may be severely limited. A completely metallic valve may be necessary to handle an extreme temperature but may not provide the shutoff required.

What is the pipe size? Is the valve being considered available in that diameter?

How fast must it operate? An automated ball valve can cycle dozens of times per minute, but a gate valve cannot.

Should the valve be able to be maintained in line? Should it be a top-entry or three-piece design?

How much should it cost? Some 1-inch (in) valves cost as much as $2000, but depending on the application they may still be considered economically feasible.

What about weight? A gate valve can weigh as much as 10 times that of a butterfly valve.

How critical is availability? Is the valve you select in stock? Can it be automated? Are repair parts available?

Throughout this book you will find the answers to these and many other questions. You may even find that there are things to consider that are unique to your needs.

Valve selection considerations

Pressure ratings. When selecting a valve for a given service, the inlet pressure and the maximum pressure drop rating of the valve must be considered. Inlet pressure ratings are generally expressed in terms of the American National Standards Institute (ANSI) class ratings and range from ANSI class 125 through 2500 depending on a valve style, size, and materials of construction including seat materials. Valves with ANSI class ratings are designed to withstand pressures greater than the nominal classifications. An ANSI class 150 valve for instance is actually rated at 285 pounds per square inch (lb/in^2) gage at 100°F. Because pressure and temperature are interactive they must both be considered in valve ratings and selection. Differential pressure, also called ΔP, is the difference between the upstream and downstream pressures of a valve in the closed position and is an important consideration in determining valve torque for actuator sizing.

According to the ANSI standard B16.5 the maximum operating pressure ratings for standard flanged- and weld-end valves must conform to the pressure and temperature limitations as shown in Table 1.1.

Temperature ratings. Temperature capabilities are essentially a function of the materials used in the construction of a valve. Temperature

TABLE 1.1 Maximum Operating Pressure Ratings for Standard Flanged-End and Standard Weld-End Gate, Plug, Ball, and Check Valves

Temperature, °F (°C)	Ratings for selected classes, lb/in^2 gage (bar)*				
	150	300	600	900	1500
−20 to 100	275	720	1440	2160	3600
(−29 to 38)	(19.0)	(49.6)	(99.3)	(148.9)	(248.2)
150 (66)	270	705	1415	2120	3540
	(18.6)	(48.6)	(97.6)	(146.0)	(244.1)
200 (93)	260	675	1350	2025	3375
	(17.9)	(46.5)	(93.1)	(139.6)	(232.7)
250 (121)	255	665	1330	1995	3325
	(17.6)	(45.8)	(91.7)	(137.5)	(229.2)

*1 bar = 100 kilopascals (kPa).

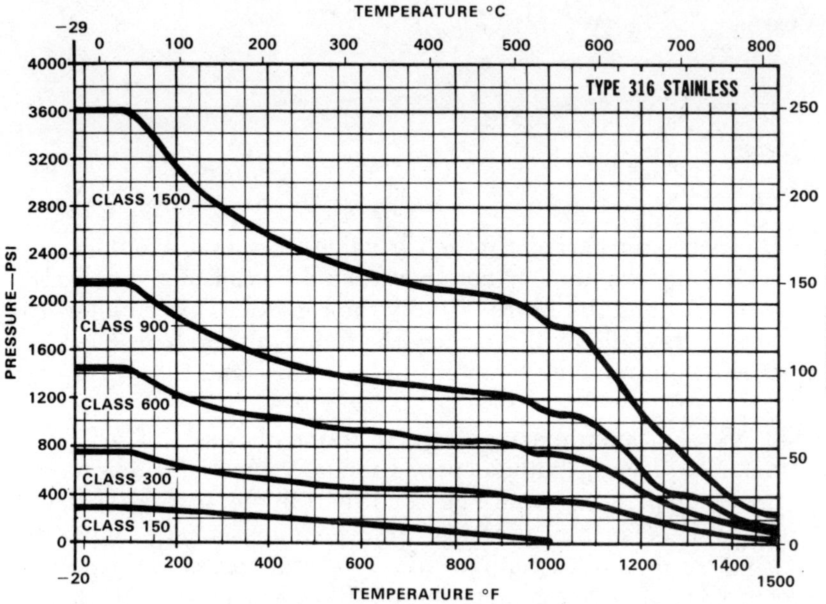

Figure 1.6 Pressure-temperature graphs aid in valve and materials selection. (*Courtesy of Fisher Controls International, Inc.*)

and pressure ratings are interactive. That is, the higher the temperature, the lower the pressure capability. For particular material combinations it is necessary to refer to the manufacturer's specifications. An example of the pressure and temperature relationship is shown in the accompanying graph (Fig. 1.6.)

Material selection. Material specification is an important step in valve selection as it has great impact on valve life, performance, and cost. Material compatibility with a given process fluid and the physical ability to withstand wear, pressure drop, and erosive fluids are basic considerations in material selection.

The fluid to be handled by a valve may be broadly categorized by

- Material
- Pressure
- Temperature
- Velocity

and subcategoried as

- Clear and clean

- Viscous
- Slurry
- Corrosive
- Erosive
- Cryogenic

How they are transferred through a valve (temperature, flow rate, etc.) determines whether or not they should be considered abrasive, erosive, or clogging in nature. Finally the conditions which allow the fluid to pass through a valve or how the fluid is prepared or used in a process will determine its temperature, pressure, and flow rating.

Shutoff. Shutoff is ordinarily stated in terms of classes of seat leakage. In actual service, shutoff depends on many factors including pressure drop, temperature, and the condition of the seating surfaces. ANSI seat leakage classifications provide a basis for comparison among valves of similar configuration. Control valves are generally not required to provide complete shutoff. Block valves are intended for complete shutoff. Some terms used to describe this are leak tight, bubble tight, or positive shutoff.

Again ANSI publishes a standard of valve leakage specifications and classes. This standard was originally adopted as FCI-70-2 (Fluid Controls Institute). The following excerpt is essentially identical for ANSI.

> 4. LEAKAGE SPECIFICATIONS AND CLASSES
>
> 4.1 The maximum allowable seat leakage as specified for each class shall not exceed the seat leakage in Table 1 test procedure as defined in section 5. For Classes II through IV each and every valve shall be tested.
>
> 4.2 Leakage Classes
>
> 4.2.1 Class I. A modification of any Class II, III, or IV where design intent is the same as the basic class, but any agreement between user and supplier. No test is required.
>
> 4.2.2 Class II. This class establishes the maximum permissible leakage generally associated with commercial double port, double-seat control valves or balance single-port control valves with a piston ring seal and metal-to-metal seats. Use test procedure Type A.
>
> 4.2.3 Class III. This class establishes the maximum permissible leakage generally associated with Class II (4.2.2), but with a higher degree of seat and seal tightness. Use test procedure Type A.
>
> 4.2.4 Class IV. This class establishes the maximum permissible leakage generally associated with commercial unbalanced single-port, single-seat control valves and balanced single-port control valves with extra tight piston rings or other sealing means and metal-to-metal seats. Use test procedure Type A.

4.2.5 Class V. This is usually specified for critical applications where the control valve may be required to be closed, without a blocking valve, for long periods of time with high differential pressure across the seating surfaces. It requires special manufacturing assembly and testing techniques. This class is generally associated with metal seat, unbalanced single-port, single seat control valves or balanced single port designs with exceptional seat and seal tightness. Use test procedure Type B using water at the maximum operating differential pressure.
4.2.6 Class VI. This class establishes the maximum permissible seat leakage generally associated with resilient seating control valves either unbalanced or balanced single port "O" rings or similar gapless seals. Use test procedure Type C.

Table 1

LEAKAGE CLASS	MAXIMUM SEAT LEAKAGE
Class I (See 4.2.1)	See Paragraph 4.2.1
Class II (See 4.2.2)	0.5% of rated valve capacity
Class III (See 4.2.3)	0.1% of rated valve capacity
Class IV (See 4.2.4)	0.01% of rated valve capacity
Class V (See 4.2.5)	5×10^{-4} ml per minute of water per inch of orifice diameter per psi differential (5×10^{-12} m^3 per second of water per mm of orifice diameter per bar differential).
Class VI (See 4.2.6)	Leakage per Paragraph 5.3.4 as expressed in ml per minute versus port diameter.

End connections. End connection selection is usually a simple question of whether the desired end connection style is available for the type of valve being considered. Rotary valves are generally available in threaded, welded, flanged, or wafer designs. Wafer-style bodies provide an initial cost savings compared to flanged designs. However, installing wafer-style valves requires careful centering between two line flanges. There are many other types of end connections available as shown in Fig. 1.7.

Cost considerations. Economics generally becomes a significant factor in valve selection. To help put economics in perspective, we can compare valve costs as a function of size for the various types of valves discussed (see Fig. 1.8). One of the challenges in making the proper valve selection is to ensure that it has a lifelong value. To do so it is important to consider the initial purchase price of the valve, any anticipated maintenance, and the expected valve life for a given application. A simple formula for calculating the "real" cost of a valve and including this criteria would be

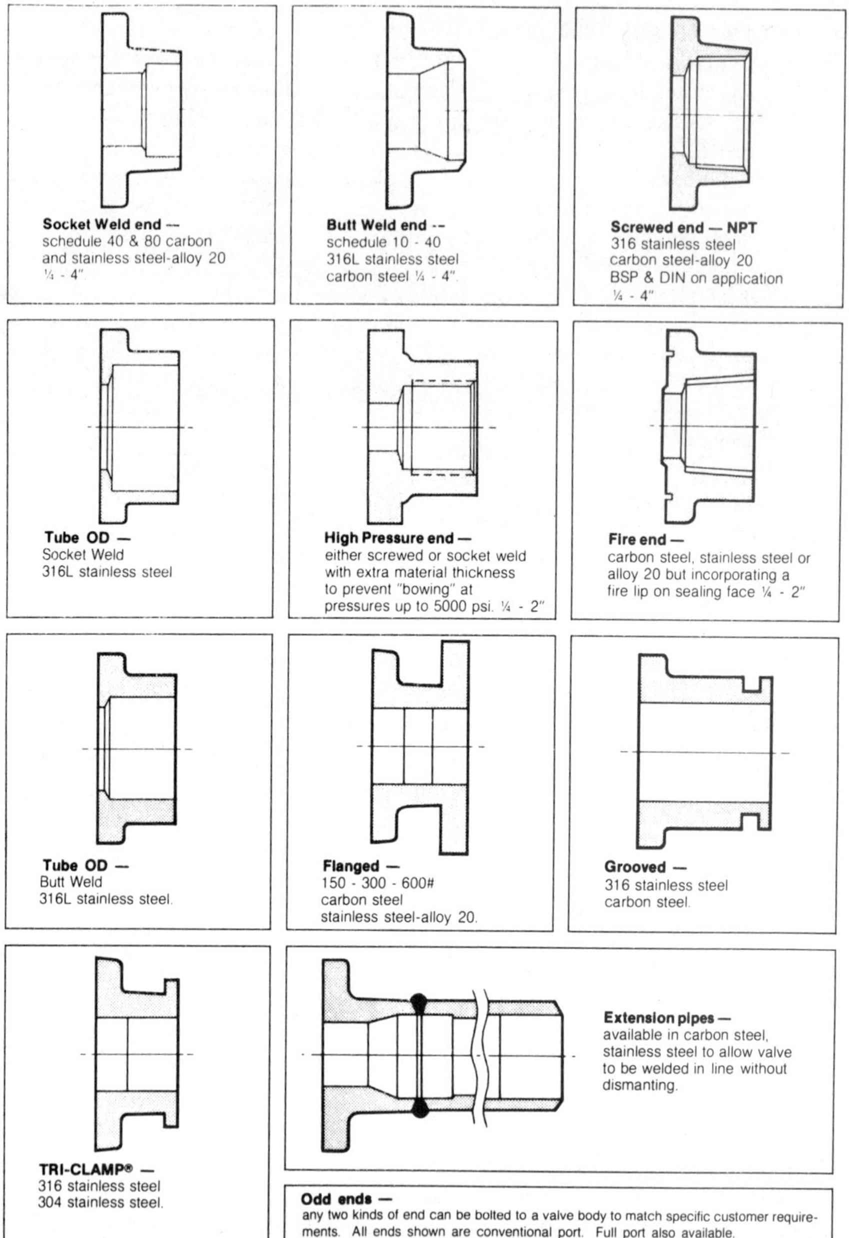

TRI-CLAMP® IS A REGISTERED TRADEMARK OF THE LADISH CO.

Figure 1.7 Valves are available with many end connections depending upon their size, style, type, and service. (*Courtesy of SVF, Inc.*)

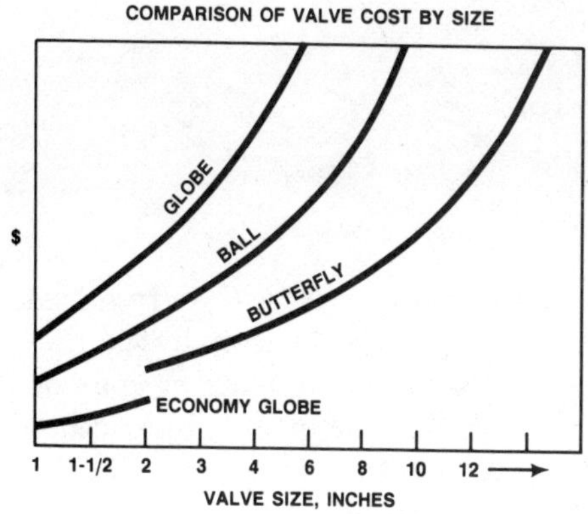

Figure 1.8 Rotary valves offer economy and high flow especially in larger sizes. (*Courtesy of Fisher Controls International, Inc.*)

$$C = P + I - S$$

where C = actual valve cost
P = purchase price
I = installation cost
S = overall savings based on proper selection

If, for instance, a valve costs \$350.00 (P = \$350.00) and takes 2 worker-hours to install (at least \$30.00 per worker-hour, I = \$60.00), the formula would now look like this:

$$C = \$350.00 + \$60.00 - S$$
$$= \$410 - S$$

Cost savings expected from a properly selected valve would include

1. Reduced maintenance
2. Simplified maintenance
3. High flow capacity
4. No leakage and no product waste
5. Reduced spare parts inventory and management

Over a year these items can result in as much as \$150.00 in savings over an inferior selection. Now the real cost of the valve may be considered to be

$$C = \$410.00 - \$150.00 = \$260.00$$

Of course this could be multiplied by the number of valves required for a given application to provide an even more realistic indication of cost. If this valve happened to require an actuator, there are other significant cost factors to consider, for example,

1. Valve torque and actuator size
2. Actuator type
3. Frequency of actuation
4. Actuator maintenance

Maintenance. Depending on the frequency and extent of required maintenance, service costs can become a significant factor in the lifetime cost of ownership of a given valve. Rotary valve packing may be inspected and maintained while the valve remains in line. Most other maintenance procedures require that the valve be removed from the pipeline. This may result in expensive down time or require parallel (redundant) piping and controls.

Another way to help reduce the cost of valves is to use remanufactured, or "used," valves. Many valve users have seen the advantages or reconditioned or rebuilt valves for reducing cost and waste. Their concern, however, is valve quality, performance, and safety. To address these issues the Valve Manufacturers Association (VMA) has established the Valve Remanufacturing Council (VRC). A report published by the VRC included a list of considerations for users planning to purchase surplus valves.

Suggestions for identifying surplus or rebuilt valves*

1. Inspect identification tags. Those that are faked may not have a part number and the stamping may be irregular.
2. Fresh sandblasted appearance including eye bolts, grease fittings, and stem.
3. Evidence of previous bolt-head scoring on backsides of flanges.
4. Bolting with excessive wrench marks on flats.
5. Direction from which the bolt head is positioned.
6. Obvious weld repairs.
7. Improper lubricant on stem nuts. Lubricant should be No. 2 grease on brass and drill collar grease on stainless.
8. Improper nut and bolt material, i.e., a bronze nut on a stainless stem.
9. Wear marks on any painted surface (handwheel, etc.).
10. Undersized dimensions (short end-to-end, flange thickness, etc.).
11. Location and attachment of identification (ID) tags. Is the ID tag on the bonnet or yoke? Is it attached with screws or rivets?

---

*Source: DuPont.

12. Less-than-quality reconditioners of valves sometimes tamper with the markings on the body casting. Be aware of casting markings that have been ground off and then other markings stamped on the ground-off area.
13. Dimensional check. Be sure dimensions are correct for the pressure class of the valve.
14. Always buy valves from known, established, reputable suppliers. Dealing with unknown strangers presents the possibility of a questionable product.
15. Question a bargain price. Any price swing of 20 percent or more should be cause to question what you are getting. Prices do not fluctuate that much unless someone is dumping inventory or substituting scrap—surplus or rebuilt valves.

Actuator requirements. Another example of the importance of the valve selection process involves actuation. Consider for instance that a ball valve and a plug valve are equally acceptable for a given service. Because they are to be automated, the next natural step is to consider the operating torque of the valve. Plug valves traditionally require much more torque than ball valves. In fact, ball and plug valve manufacturers' published torque often differs significantly for the same sizes. This means that an actuator may be physically larger and more costly for a plug valve than for a ball valve. Other considerations are noise, created as the fluid rushes through the tortuous path of some on-off valves or a modulating valve; cavitation; flashing; and even the speed of operation.

Summary. Mixing and matching these many characteristics makes the task of valve design and selection monumental. In fact, the reason that there are so many valve types is that there are so many different requirements from the users of valves.

1.2 Basics of Valve Design (Seats and Seals)

A valve is properly considered to be a hybrid structure—pressure vessel and operating machinery. The important performance parameters of a valve are pressure-boundary integrity, operating effort, and control and/or shutoff performance.

Not all valves can handle the pressure and temperature ratings of all applications, but pressure and temperature are two basic properties of a fluid that must be considered in valve selection. The ANSI and Manufacturers Standardization Society (MSS) establish pressure and temperature ratings for valves. They are based on the design and materials of construction of the valve. Valves used in industrial applications should meet the quality standards recommended by one or more national and/or international standards organizations. Quality control encompasses design, materials, inspection, test, and usage limitations. There are several codes, regulations, and standards that apply to large segments of the valve industry.

A process fluid must be fully contained in a properly designed piping system. In a pipeline there are many potential leak paths for a fluid: pipe joints, seams, equipment connections, and valves. To understand valve design as it relates to containing pressure, it is best to look at the basic components. The body is the main pressure-containing element of a valve. It also serves as the basic foundation for all the other components including the end connections. The valve body must also withstand loads generated by piping connections and the actuator. There are many considerations that must go into the design of a valve body including valve type, pressure rating, manufacturing, material, and cost. Depending on the type of valve, there may be a variety of potential leak paths through the body. There is the bonnet attachment, body seals, and stem seal as well as the pipe connection area.

Pressure-boundary integrity

Pressure-boundary integrity requires basically sound pressure parts, pressure-tight static assembly joints, and, in most cases, an effective working seal between a moving stem and the valve bonnet.

Sealing the valve stem. Because a valve is intended to contain the fluid that it is controlling, the basic design to meet this requirement is complicated by the need for a valve stem which penetrates the body. A very important element then for complete pressure containment of a valve is the stem seal. To contain the fluid and allow the stem to operate, valve manufacturers are continually working with new designs and materials to seal the stem area. The stem seal is the most common site for fluid leakage through a valve because it is a dynamic seal.

In terms of stem-packing leaks, all rotary-action valves offer a significant improvement over sliding stem linear valves. Sealing the stem of a gate or globe valve is usually accomplished by filling a cylindrical chamber in the bonnet surrounding the stem with a packing material compressed in the chamber by a gland and associated bolting (see Fig. 1.9). In a standard globe valve, the packing acts as a wiper as the stem goes up and down. A difficult-to-hold gas, such as ammonia or chlorine, can migrate through the packing to mix with the atmosphere or cause corrosion of the stem. The alternating wet and dry stem as it cycles can increase corrosion and leakage and damage packing and thus cause a more significant leak. Another concern is in abrasive-slurry and high-

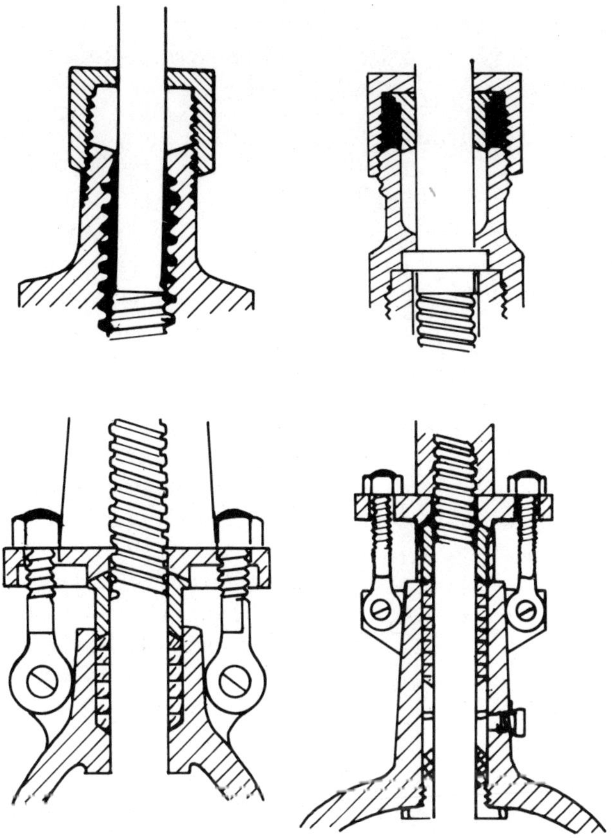

Figure 1.9 Packing material is compressed in the packing chamber to create a stem seal. (*Courtesy of Crane Company.*)

temperature service. Continual up and down action can again create packing damage, causing a leak to develop. The reciprocating action of the stem against the packing in a gate or globe valve is not present in rotary valve designs, and so they tend to seal longer.

Automated quarter-turn valves are subject to premature stem seal failure when there is misalignment between the actuator drive shaft and the valve stem. This is usually caused by improperly manufactured mounting hardware. Careful selection of valve and actuator suppliers can help ensure high-quality materials and well-designed mounting hardware.

Stem seal leakage usually results from packing wear and can often be corrected by tightening the stem packing gland. Excessive tightening can cause high stem friction, difficult valve operation, and accelerated wear. This would mean, of course, that the operating torque of an automated valve would increase and might even stall the actuator.

If an automated valve is in a critical service and cannot be removed for immediate maintenance, the actuator should be sized for an operating force assuming maximum packing tightness.

Valve seats. Improperly selected valve seat materials is the most common reason for internal leakage. To select the proper seat it is necessary to generate a complete fluid profile. This includes pressure and temperature but, more importantly, a classification of the fluid such as clean liquid, gas, steam, viscous, fibrous, or powder. This helps the valve manufacturer to make a seat material recommendation. Additionally, valve operation may be affected by the fluid class. Valve manufacturers often apply actuator sizing factors or increase the valve operating torque. Use of a valve in severe throttling conditions can be harmful to valve seating surfaces and materials, and in cases where the fluid is cavitating, gross damage to the valve body and/or downstream piping may result.

Valves which remain in one position for long periods of time may be subject to some degree of reduction of operability as a result of loss of effective lubricants in threads, aging of packing, surface corrosion of moving parts, accumulation of deleterious solids, or material cold flow. In some applications it may be desirable to schedule periodic partial- or full-cycle exercising of such valves. This may be done with manual valves by cycling them periodically. When automated valves are to be exercised regularly, provisions must be made to do so. An example would be a manual override device on the pilot valve of a pneumatic actuator.

Stem seal designs for severe service valves

A fluid leaking through any part of a closed piping system is referred to as a fugitive emission. Valves, because of their sheer numbers, are the major contributor to overall fugitive emission totals in the process industries. The variables affecting emission rates are valve types and valve packing configurations.

Environmental concerns. Asbestos materials formerly used in valve construction for nonmetallic sealing components—gaskets, packing, and soft seats—have been replaced by other materials, including polytetrafluoroethylene (PTFE), encapsulated fiber asbestos, and graphite laminate and filament. But as with many other manufactured items, the task of identifying suitable replacements for asbestos materials in valves is ongoing.

New Occupational Safety and Health Administration (OSHA) regulations, Superfund Amendment Reauthorization Act (SARA) Title III, and amendments currently being considered for the Clean Air Act establish stringent guidelines for control of fugitive emissions of many chemical substances, the most prominent of which is benzene. Because of the high pressures and temperatures encountered in piping systems, the regulations impact heavily on valve manufacturers and users. The need for zero tolerance in valve stem leakage is a hot topic.

The current practice for stem seals includes the use of stuffing boxes with graphite packing, fluoropolymer seals, and packless valves using diaphragms or bellows, for applications demanding tight specifications. Because valves are the major source of fugitive emissions in the process industry, the types of valves installed in chemical plants and refineries should be considered.

A recent study by the Environmental Protection Agency (EPA) revealed the following usage practices in two major industries.

Industry Valve Installation Practice (process service only)

Valve type	Refineries	Chemical plants
Gate	Great majority	Few (phasing out)
Unlined plug	Small population	Essentially none
Nonlubricated plug	None	Major population group
Ball	Few	Small to large quantities
Butterfly	Rare	Small population
Globe	Small population	Probably less than 20%

Leak-frequency data for valves in a chemical plant obtained by the EPA indicate that gate valve leak frequency is 10 times greater than lined plug valve leak frequency. The reasons for lower lined plug valve emission rates may be due to the continuous downward pressure and close fit of the plug and sleeve accompanied by the long path length from the top of the plug to the fluid opening in the plug. In contrast, unlined metal plug valves frequently are lubricated, which tends to increase the wall clearance around the plug and, consequently, the fugitive emissions rate.

In the last decade improvement in valve construction and valve packing has had a major impact on valve emissions, particularly in gate valves. Currently, a wide variety of packings that are graphite- or Teflon-based have become substitutes for asbestos. Packings based on Teflon have been used for a number of years on quarter-turn valves. Studies indicate that the graphite emission leakage rate is about 10 percent of the asbestos packing emission rate. Emission control with graphite packings, however, requires specific packing configurations and graphite physical characteristics for optimum emission control. While emission data for other packings are lacking, it is believed that the use of most asbestos substitutes will result in lower emission rates than if asbestos-packed valves are used.

Today's plant personnel must not only concern themselves with the type of packing material to put into their valves but with the sealing effectiveness of the packing technique. One option is to employ the live-loading technique. This concept, originally used to pack valves in nuclear service, is now beginning to take hold in every type of industry.

Severe duty quarter-turn valves are one of the workhorses of the chemical process industry (CPI). Although there are varying definitions of severe services and the kinds of valves that handle those applications, severe service quarter-turn valves are mostly used for processing extremely hazardous materials (phosgene, toluene, chlorine, vinyl chloride monomer, nitric acid, hydrofluoric acid, acetic acid, and zinc sulfate, to name a few) or where expensive or exotic fluids are being handled.

Severe service valves differ from conventional ball or plug valves in two fundamental ways.

1. The stem packing area is longer. In some designs there is as much as 10 times more packing material. This is referred to as redundant seal technology. Many have a "sniffer" port at the midpoint between the packing so that leakage past the first series of packing can be detected and corrected before any of the contained material leaks to the atmosphere (see Fig. 1.10).

2. The other difference is in the loading of the packing area. Con-

Figure 1.10 Severe service valves have extra packing that is spring loaded and a sniffer port to detect leakage. *(Courtesy of Neles-Jamesbury, Inc.)*

ventional valves have a gland nut or some other method for compressing the packing to effect a seal. If leakage occurs, the gland nut may be retightened. Severe service valves use a live-loading method. Here spring-loaded packing or packings are loaded with belleville washers on the gland followers. This provides for a degree of self-adjustment to compensate for wear. A severe service valve is typically less expensive than an equivalent bellows sealed valve (see the following section on Leakless Valves).

Live loading of valve packing is accomplished by storing energy in spring washers (belleville). This technique puts a compressive force on the packing and maintains a relatively leak-tight seal around the stem while allowing the stem to move as necessary during opening and closing.

The number of springs varies according to the valve configuration, but it usually is possible to design a group of springs that maintains about 80 percent of the original design load, even when the packing has seen some consolidation while in service.

A reduction in the number of packing rings in a valve is one of the benefits of live loading. Too many rings results in increased breakout friction, high operating torque, increased running friction, and, sometimes, inoperable valves. Packing material should remain flexible throughout the life of the valve. After initial torque, the packing—especially in high temperature applications—will relax radially and must be manually adjusted to effectively seal. With live loading, seal adjustment is done automatically.

Properly designed live-loaded valve packing improves the control of emissions leakage; increases reliability of packing, providing an additional safety factor in areas not accessible during normal operations; and saves on packing by increasing its life expectancy and decreasing total erosion.

The use of springs to live-load a valve also provides a guide to help the user determine if there is sufficient bolt torque on the packing. The visual flat position of the spring will indicate that the springs are fully loaded. The visual position of the spring being opened indicates torque loss or insufficient torque loading on the packing. Any valve that has a packed stem, sufficient clearance, and a gland stud long enough to hold the springs can be live-loaded. The springs produce a constant force against the packing that can be more than 20 times greater than could be applied manually with a gland nut. Even with the use of only one spring, the stored energy available to maintain the force on the packing is 4 times greater than the manual force normally used on the gland nut to maintain packing load.

The following types of valve situations would benefit from the enhanced packing design used with severe service valves.

- Valves that are subjected to a high degree of thermal cycling
- Motor-operated, air-operated, and manually operated valves that are frequently actuated
- Critical valves with a possible high volatility
- Certain types of valves that have been particularly troublesome from past experience

Leakless valves

There are basically three valve designs that to some extent can be considered leakless: i.e., these valve types provide a 100 percent bonnet seal which eliminates leakage around the valve stem. The three leakless valve types are diaphragm valves, pinch valves, and sealed bellows valves.

However, *zero leakage* is the new buzzword when it comes to valves being used in the CPI. The valve most often mentioned to accomplish this is the bellows-sealed valve (see Fig. 1.11), which provides a secondary seal such as a membrane or gland in the stem. A sliding-stem-type control valve designed for the elimination of fugitive emissions uses a bellow seal assembly with backup packing and an analyzer port. A hydroforming process shapes the corrugations into the flat bellows material before the tube is formed, resulting in less stress at the crests of the corrugations and making the bellows more resilient. The installed bellows neutral state is with valve stroke at normal, half-open operating position. The bellows is 10 times the length of the normal operating stroke of the valve. An antirotation lock prevents twisting and premature bellows failure. A monitoring port allows for remote sensing of bellows failure. A secondary seal located above the test chamber prevents leakage in the event of bellows failure. Bellows-sealed valves are available for ANSI classes 150, 300, 600, and 1500, and in sizes from ½ to 24 in [13 to 600 millimeters (mm)]. They are available in cryogenic, angled, jacketed, and other special configurations.

With bellows-sealed valves, "packless" technology extends beyond smaller diaphragm valves to much larger gate and globe valves. Bellows seals have been used successfully for many years in gate and globe valves and are also available in quarter-turn valves. Bellows for quarter-turn valves are available from ½ to 6 in (13 to 150 mm) in class 150 or 300 full or reduced port ball valves (see Fig. 1.12). Bellows are available in stainless steel, Inconel, Monel, and tetrafluorethylene materials.

A maintenance-free valve such as the bellows seal may cost anywhere from 1½ to 5 times more than another valve used in the same

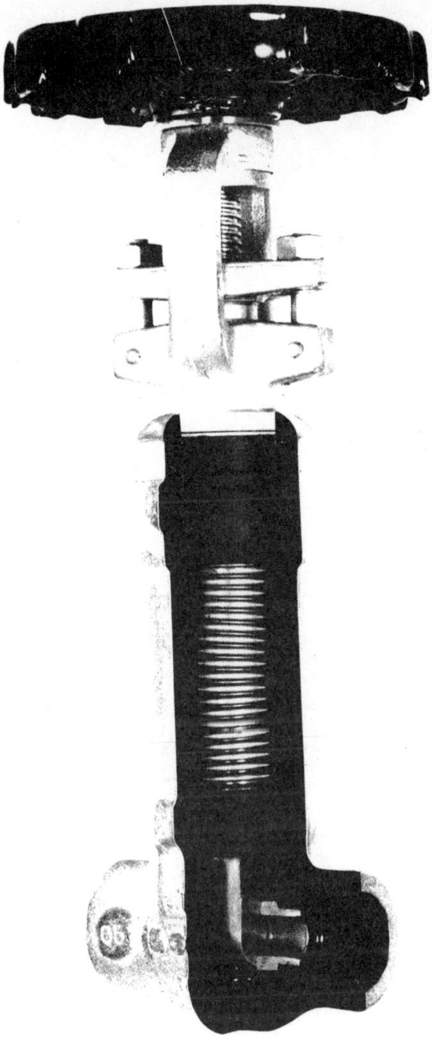

Figure 1.11 Bellows-seal valves are considered leakless. (*Courtesy of Kerotest Mfg. Corporation.*)

application. But if one considers the cost of a monitoring system, plant shutdown, EPA fines for emissions, workmen's compensation, cleanup, public relations to the community, and high insurance premiums, the cost may be justified.

1.3 Materials Used in Valve Construction

A valve can achieve no more than its material will allow. With each new development in materials or their properties the ability of valves

Figure 1.12 Bellows technology is used to eliminate external stem leakage on rotary valves. (*Courtesy of Kerotest Mfg. Corporation.*)

to handle more severe applications is improved. Typically, valves are constructed of two types of materials that may be referred to as *hard* and *soft*. All the components that contain the fluid pressure make up the valve body. The body and the trim (ball, stem, gate, etc.) are constructed of hard materials. Hard materials must withstand the mechanical stresses of fluid pressure, abrasion, bolting, threading, weight, and piping stresses as well as the corrosive condition of not only the surrounding environment but certainly the fluid that it contains.

Soft materials are those whose properties make them useful as seats, seals, and liners. Various compositions of fluorocarbons, fibers, graphites, and elastomers provide for an incredible selection of material properties to satisfy most applications.

Material selection is often quite simple for nonaggressive fluids. But even water must be carefully analyzed before a material is selected. For instance, stainless steel type 17-4 pH is corrosion resistant. It will not corrode in ordinary tap water, but it will corrode in pure demineralized water. Titanium is dangerous for dry chlorine but excellent for wet chlorine. But the opposite is true for copper.

Each combination of body, trim, and seal materials may provide desirable performance on one hand but undesirable performance on the other. For instance, where desirable corrosion resistance is achieved, undesirable pressure or temperature limitations may exist. When cost is a consideration, one may accept a stiff purchase price for the offsetting longer performance life of a specific valve. Fortunately there are plenty of materials from which to choose.

Valve bodies and closure elements may be constructed of just about any material that may be cast, forged, molded, or welded. This includes iron, steel, aluminum, bronze, nickel, titanium, and many other metals and alloys. They may also be made from a variety of plastics and even ceramics. Some valve types and sizes are not available in all materials. In some cases a corrosion-resistant metal may prove inadequate or too costly for a particular service. Oftentimes a plastic or elastomer lining may be available. The strength of an iron or steel body in combination with the unique properties of the liner material is sometimes an excellent cost-saving alternative. Some examples of these valves are diaphragm valves and fully lined ball, plug, and butterfly valves. Eccentric plug valves often feature an elastomer-coated plug. With lined equipment the liner material must serve as the critical barrier between the corrosive fluid and the surrounding environment.

It pays to know the range of materials from which valves are usually made and to understand the pressure, temperature, and structural limitations of each material. It may be highly unsafe to use materials for services they are not recommended for.

Metals used in valve construction

Bronze. Bronze is widely used in gate valves and fittings for temperatures up to 450°F (230°C). It is an alloy of copper, tin, lead, and zinc. As a ball valve material its usable temperature is limited by the seat material.

Iron. Cast iron is regularly made in three grades—cast iron, ferrosteel, and high tensile iron. These metals are recommended for temperatures up to 450°F (230°C). Fully lined valves often use the structural strength of iron to support plastic or elastomer lining materials.

Steel. Steel is recommended for high pressures and temperatures and for services where working conditions, either internal or external, may be too severe for bronze or iron. Its superior strength and toughness and its resistance to piping strains, vibration, shock, low temperature, and damage by fire afford reliable protection when safety and utility are desired. Many different types of steel—cast, forged, alloy—are both necessary and available because of the widely diversified services steel valves perform.

Stainless steel. Stainless-steel castings should be heat-treated for maximum corrosion resistance, high strength, and good wearing properties. Seating surfaces, stems, and discs of stainless steel are well suited for severe services where foreign materials in the fluids handled could have adverse effects. Stainless steel is resistant to wear, seizure, galling erosion, and oxidation.

Carbon and alloy characteristics. The following tables show several alloys and the percentages of their chemical components.

WCB Cast Carbon Steel (ASTM A-216 Grade WCB)

Chemical component	Percent of component
Carbon	0.30 max.
Manganese	1.00 max.
Nickel	0.50 max.
Chromium	0.50 max.
Molybdenum	0.20 max.
Copper	0.30 max.
Silicon	0.60 max.

WCB cast carbon is an excellent metal for pressure-containing parts used in high-temperature service. Its strength and toughness assure high resistance to shock, vibration, piping strains, and fire and freezing hazards. Limiting the carbon content to below 0.30 percent provides good weldability. It is suitable for saturated or superheated steam, hot or cold water, and hot or cold noncorrosive oils, fats, air, or other fluids. Consideration should be given to the possibility of graphitization above the rated service temperature range.

CF8M (type 316) Stainless Steel (ASTM A744)

Chemical component	Percent of component
Carbon	0.08 max.
Manganese	1.50 max.
Silicon	2.00 max.
Chromium	18.0–21.00
Nickel	9.00–12.00
Molybdenum	2.00–3.00

The addition of molybdenum to the 18 percent chromium, 8 percent nickel base alloy greatly widens its field of usefulness, so it has become the most widely used alloy in the construction of corrosion-resistant valves. The presence of molybdenum greatly increases resistance to attack by reducing agents and lessens the tendency toward pitting. The alloy is widely used in the sulfite and sulfate pulp industry, in the manufacture of most organic acids and their derivatives, and in the textile fibers industry in handling the many intermediate chemicals usually encountered.

CF8 (type 304) Stainless Steel (ASTM A744)

Chemical component	Percent of component
Carbon	0.08 max.
Manganese	1.50 max.
Silicon	2.00 max.
Chromium	18.0–21.00
Nickel	8.00–11.00

This alloy is widely used in the construction of valves. The chromium produces passivity by formation of an impervious film which is resistant to oxidizing solutions. Nickel produces the austenitic structure and contributes to the corrosion resistance. This alloy is also widely used in the cryogenic services and in services where freedom from contamination is essential. The alloy is not suited for handling reducing agents or sulfuric or hydrochloric acids.

CN7M (FA-20) (ASTM A744)

Chemical component	Percent of component
Carbon	0.07 max.
Manganese	1.50 max.
Silicon	1.50 max.
Chromium	19.0–22.00
Nickel	27.50–30.50
Molybdenum	2.00 3.00
Copper	3.00–4.00

CN7M provides excellent corrosion resistance to a wide range of sulfuric acid concentrations at elevated temperatures. Because the chromium provides resistance to oxidizing solutions, it is used in many applications where CF8 and CF8M are not wholly satisfactory. CN7M has a wide application in handling caustic soda, organic acids, chlorinated hydrocarbons, sludge acids, and many other process solutions.

Preventing material failure. To protect against stress corrosion cracking (SCC), a progressive type of material failure caused by surface tensile stress plus a concentration of ions in the attacking fluid, valve manufacturers have tightened specifications for selection, fabrication, and heat treating of metals used in valve construction. These include certain types of precipitation-hardened stainless steels and of nickel-based alloys such as Monels, Inconels, and Hastelloys which are highly resistant to SCC in most environments.

Besides using American Society of Tool and Manufacturing Engineers (ASTM) and American Society of Mechanical Engineers (ASME) specifications, one major valve manufacturer uses argon-oxygen-decarburization (AOD) refining, a relatively new foundry process that provides close control critical for high-nickel alloy castings. AOD refining ensures uniform product, necessary because even minute impurities can lead to premature failure of valve components. The resultant valve, with body, bonnet, plug, stem, seat ring, and seat ring retainer constructed of high-nickel alloys, allows maximum inlet pressures of 1480 lb/in^2 gage (102 bar) and process temperature capabilities ranging from −325 to 800°F (−200 to 425°C) depending upon specific construction and pressure requirements.

How long a material, hard or soft, will last depends on many factors. Pressure, temperature, service selection, material quality, and how the material is used in the valve all must be considered in addition to the fluid characteristics.

Nonmetallic materials

Even though metals will continue to be used throughout the valve industry for their own unique characteristics of cost and performance, it has been the many developments in plastic and other nonmetals that have greatly expanded the applications of valves. The following is a brief review of some of the nonmetallic materials being used in today's valve designs.

Fluorocarbons are used extensively in chemical process equipment because of their chemical inertness. Valve seats, liners, and in some cases entire valves may be constructed of these materials. Manufacturers choose specific fluorocarbons [tetrafluorethylene (TFE), fluorinated ethylene-propylene (FEP), or ultrahigh-molecular-weight polyethylene (UHMWPE)] based on the comparative resistance of these plastics to permeation and absorption of chemicals in addition to their physical and mechanical properties.

TFE. The properties of TFE which make it particularly useful in valves for chemical process are lubricity, nonpermeability, nonstick characteristic, abrasion resistance, corrosion resistance, etc., which can be realized at temperatures below 450°F (230°C). To make the best use of these properties, valves containing TFE materials are generally restricted to temperatures below 400°F (200°C).

TFE deforms under pressure or temperature or over a period of time; however, it will always tend to return to its original shape. This deformation is known as *cold flow*. When you remove either the pressure or the temperature, the inherent memory of TFE will tend to bring it back to its original shape. However, it does not have perfect memory as many elastomers and may return to only about 60 percent of its original shape. For this reason TFE and other fluoropolymers when used as valve seats are often backed by a resilient material or spring. When continuous pressure on the seat as on a plug valve is employed, the cold flow can cause a mechanical interlocking of the closure element and seat which increases operating torque.

TFE is here used as a seal for rotary-motion valves and for some linear valves. It also is used as a liner and an encapsulant for ball, butterfly, and plug valves (see Fig. 1.13). It is attacked only by molten alkali metals, or by chlorine or fluorine trifluoride under special conditions. So corrosion is not a problem. It is a plastic, not an elastomer. When deformed it will partially recover slowly. It does not resist abrasion well and should not be exposed to highly abrasive streams. The nominal temperature range of TFE is from cryogenic to 400°F (200°C). When considering TFE, there are two characteristics to keep in mind. Permeation and absorption are quite similar phenomena except that permeation refers to the passage of foreign material through the polymer while absorption refers to the retention of material within the polymer. All polymers will absorb or be permeated to some degree by

Figure 1.13 Fluorocarbons are often used as a lining and encapsulant for valves as an alternative to exotic alloys. (*Courtesy of Posacon, Inc.*)

materials which they come in contact with, but fluorocarbons are more resistant to these effects than other plastics. The mechanism for the occurrence of this phenomenon is diffusion or movement of the gas or liquid through submicroscopic, intermolecular spaces within the polymer. In PTFE and FEP this is purely a physical process since these fluorocarbons are inert to chemical attack in virtually all chemicals. Temperature has an extremely important influence on the permeation or absorption of all polymers. As the temperature increases, the expansion of the polymer will increase the size of the intermolecular spaces and allow less restriction to the migration of the vapor or liquid being handled.

FEP. FEP is a copolymer of TFE and hexafluoropropylene, and it is a true thermoplastic. It is melt processible at a temperature of 634°F (335°C), making it an excellent resin for transfer-extrusion molding of parts with difficult shapes. Through precise control of the molding process, the most desirable properties of FEP—chemical inertness, low coefficient of friction, insolubility in solvents, low adhesion properties, wide service temperature range, toughness, and flexibility—can be realized in the finished valve product. Because FEP is molded and machined to close tolerances, parts such as the body and plug or liner and discs can be fitted for excellent sealing.

Perfluoroalkoxy resin. PFA is a class of perfluoropolymers that offers the processing ease of conventional thermoplastics but substantially extends its temperature limits. Like FEP, PFA is a true thermoplastic and is melt processible, allowing it to be molded to extremely difficult shapes. PFA is processed at 700°F (370°C). Otherwise the molding technology is the same as that used for FEP.

PFA has been found to be better in handling some monomers, such as butadiene, permitting the use of PFA-lined products on a wider range of applications to temperatures as high as 500°F (260°C). Where lined valves are used in abrasive service many manufacturers recommend that UHMWPE be used. UHMWPE is one of the most abrasion-resistant liner materials available today: almost 5 times as abrasion resistant as TFE and more than 6 times that of carbon steel.

Tefzel. Tefzel can best be described as a rugged thermoplastic with an outstanding balance of properties. It has mechanical toughness, broad thermal capability, and the ability to meet severe environmental conditions.

Chemically, Tefzel is a copolymer of 25 percent ethylene and 75 percent TFE. Mechanically, Tefzel is tough, exhibits high tensile strength, and is more creep resistant than Teflon, TFE, FEP, and PFA fluorocarbon resins. The Tefzel used in some valves is reinforced with glass, yielding a tensile strength approaching 12,000 lb/in^2 (830 bar). Tefzel has outstanding resistance to attack by chemicals and solvents that often cause rapid deterioration of other plastic materials. Tefzel is inert to strong mineral acids, inorganic bases, halogens, and metal salt solutions. Carboxylic acids, anhydrides, aromatic and aliphatic hydrocarbons, and classic polymer solvents have little effect on the material.

Thermoplastic valves. Thermoplastic valves provide a dependable and economical means for fluid control and management in a variety of piping systems. These valves are made entirely of thermoplastics (see Fig. 1.14) including the body, stem, and closure element.

Polyvinyl chloride (PVC). The most frequently specified thermoplastic material, PVC, has excellent strength, rigidity, modulus of elasticity, and chemical resistance. It is used extensively in water lines, irrigation, plating, chemical drainage, and chemical processing. PVC has good physical properties and resistance to corrosive and chemical attack. It is, however, subject to attack by aromatics, chlorinated or-

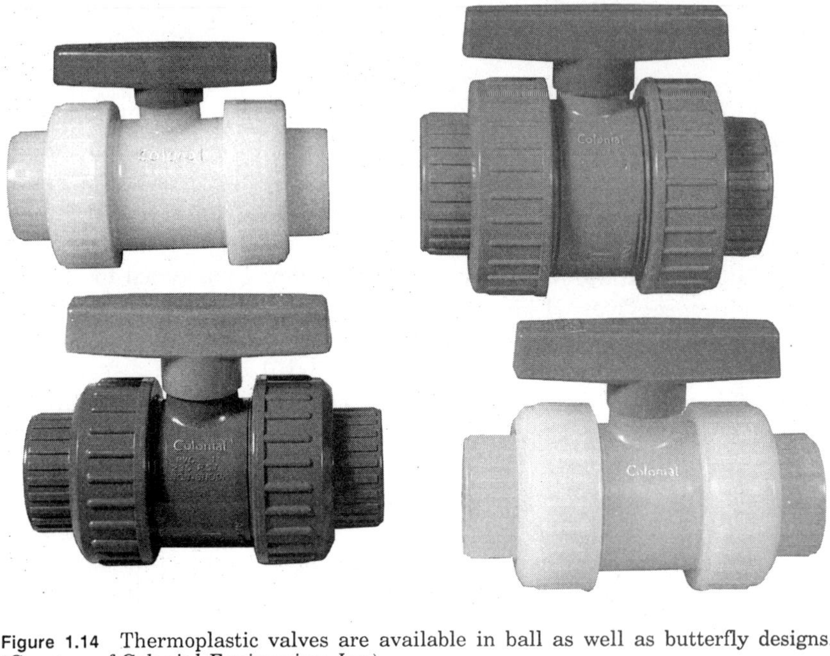

Figure 1.14 Thermoplastic valves are available in ball as well as butterfly designs. *(Courtesy of Colonial Engineering, Inc.)*

ganic compounds, some hydrocarbons, polar solvents, and ketones. PVC has a maximum operating temperature of 140°F (60°C) and has a design stress of 2000 lb/in^2 (140 bar) at 73°F (23°C). Joining is accomplished by solvent cementing, threading, or flanging.

Chlorinated polyvinyl chloride (CPVC). The physical properties of CPVC are equal to or better than PVC, and its use is ideal for hot or cold corrosive liquids, hot or chilled water, and generally where escalated temperatures preclude the use of PVC. CPVC has a maximum operating temperature of 200°F (90°C) and a design stress of 2000 lb/in^2 at 73°F (140 bar at 23°C). Joining is accomplished by solvent cementing, threading, and flanging.

Polypropylene (PP). PP is generally lower in physical properties as compared to PVC. It is chemically resistant to acids, organic solvents, and alkalines. PP is excellent for sulfur-bearing compounds and saline solutions, and its use is ideal for drainage where mixtures of solvents, base chemicals, and acids are transported. It has a maximum operating temperature of 180°F (80°C) for drainage purposes and a design stress of 2000 lb/in^2 at 73°F (140 bar at 23°C). Joining is accomplished by solvent cementing, threading, and flanging.

Polyvinylidene fluoride (PVDF). PVDF is superior in physical properties to other thermoplastic piping system components. It retains most of its strength to working temperatures as high as 260°F (125°C) and is chemically resistant to most acids, bases, solvents, bromines, chlorines, aliphatics, aromatics, and alcohols. PVDF is generally not recommended for ketones or esters. Joining is accomplished by thermoseal fusion, threading, or flanging.

Composite materials

Composite ball and butterfly valves are primarily selected for chemical handling services where internal and external corrosive conditions exist, particularly when metal or lined metal valves have proven unsuitable for the chemical service (see Fig. 1.15).

Typical materials used in molding composite valves are highly chemical and temperature resistant rigid polymers reinforced with carbon graphite or glass fibers. Polymers that have been used successfully are vinyl esters and polyphenylene sulfide (PPS). Both of these resins, which have a fiber reinforcement content in the range of 25 to 50 percent, have been used to mold valve bodies and discs for design service to withstand pressures to 275 lb/in^2 (19 bar) and fluid temperatures 350°F (175°C).

Typically, these materials offer high rigidity at both low and elevated temperatures even 50 percent higher than the rated tempera-

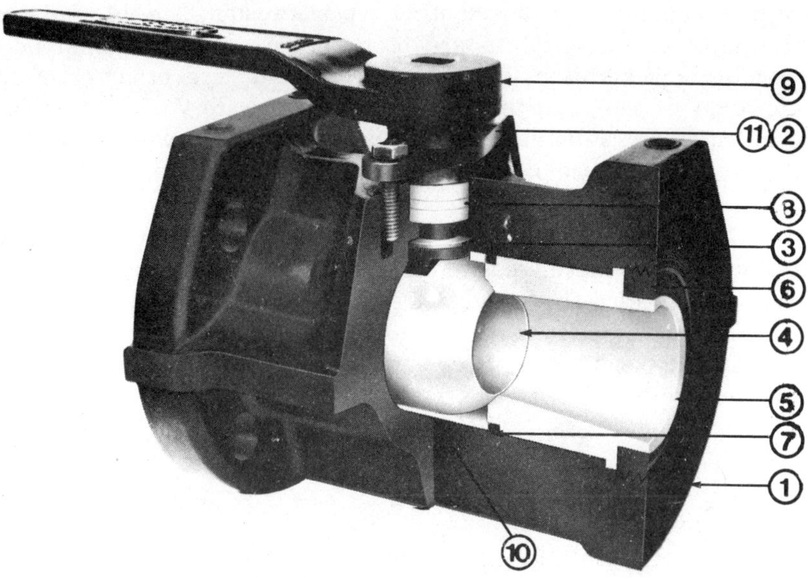

1. **Valve Body:** Graphite fiber-reinforced vinyl ester. Optional reinforced resin includes Nil-Cor 500.
2. **Gland and Bolts:** Hastelloy "C".
3. **Thrust Washer:** Graphite fiber-reinforced TFE.
4. **Ball:** Ceramic.
5. **Seat Inserts:** Ceramic.
6. **Retainers:** Kynar.
7. **O-rings:** Viton.
8. **Stem Packing:** TFE.
9. **Handle:** Composite.
10. **Body Liner:** Ceramic.
11. **Stem:** Hastelloy "C" insert in molded composite.

Figure 1.15 Composite valves are used when internal and external corrosive conditions exist. (*Courtesy of Dresser/Nilcor.*)

ture. This stability extends the valve life by providing dimensional accuracy (necessary for sealing surfaces) over extreme limits of temperature. In general, carbon graphite fiber reinforcement is used in the higher-temperature, high-chemical-concentration services because of its chemical inertness. It is also a good choice in valve components that move against sealing surfaces because of its inherent lubricity and improved surface integrity.

Valve components (e.g., valve shafts) can be molded with metal inserts to improve the physical strength of critical components. In such cases, materials within the valve body (wetted surfaces) should be fully encapsulated by the molded composite material or a metal alloy selected to give equal or better chemical corrosion resistance according to the service conditions.

Glass-reinforced polysulfone (GRP). GRP-polysulfone composite valves are intended for extremely reactive chloride-type radical compounds that have chemically attacked other types of resins. The chloride radical compounds referred to are of the type found in unbalanced hypochlorite generation, chlorine cell effluent, and other chloride-alkali-type processes. At this point in time, laboratory tests show that the modified polysulfone composite is resistant to a wide variety of corrosive transient conditions involving chloride compounds for temperatures up to 350°F (175°C).

Summary. Since thermoplastic and composite valves usually weigh less than one-third the weight of equivalently rated metal or lined metal valves, there is reduced stress on a piping system; therefore, fewer pipe supports are required. One of the major benefits of composite valves is external corrosion resistance in corrosive environments.

Other seat and seal materials used in valves

Butyl. Butyl is a vulcanized rubber copolymer generated from isobutylene and isoprene. Butyl has good chemical resistance to most acids.

Ethylene propylene diene monomer (EPDM). EPDM is used for O-ring seals and valve seats. It has excellent abrasion and tear resistance and is chemically resistant to a large variety of acids, alkalines, alcohols, and oxidants. It is recommended for use with petroleum products, strong acids, or strong alkalines.

Viton (a fluorocarbon elastomer). Viton is used for O-ring seals and valve seats. It has resistance to a broad spectrum of chemicals with greatly expanded temperature ranges. Viton is abrasion resistant and tough and resists most mineral acids, salt solutions, chlorinated hydrocarbons, and petroleum oils.

Ceramic construction of valve components

Alumina ceramic has a hardness near that of a diamond and is far superior to metals and stellited materials in wear resistance. Alumina ceramics are inert to oxidation, are not corroded by chemical agents, and are not subject to radiation damage.

Ceramic materials not only provide excellent abrasion and corrosion resistance but are also addressing higher-temperature valve applications. Some of the ceramic materials such as silicon nitride are so stable at elevated temperatures that they can be used with molten metals. Ceramic materials and their use in control valves is a rapidly

Figure 1.16 Ceramic-lined valves provide excellent service in abrasive and corrosive applications. (*Courtesy of Fujikin of America, Inc.*)

growing technology that will provide alternatives for the control valve requirements of the future and the solutions for many of the control valve problems of the present (see Fig. 1.16). Along with these advantages, fine ceramics also have some disadvantages such as brittleness, low impact resistance, low tensile strength, and susceptibility to thermal shock damage.

What to consider when using a materials selection guide

Most valve manufacturers publish in-house material selection guides (see Appendix A), but as complete as they attempt to be there are still many conditions that can complicate the selection process. Whenever possible, testing should be performed along with a healthy dose of experience.

When selecting a valve material, many variables enter into the question of serviceability. Factors such as concentration, temperature, pressure, velocity, percent solids, temperature cycling, vacuum, and cleaning practices are all important in determining whether or not a particular material will give satisfactory service.

Of the endless number of chemical compounds many are insoluble in water and would consequently cause no corrosion problems when exposed to water. However, some of these simple services can become difficult when it is necessary to make such materials soluble through use of some other solvent. For example, sulfuric acid is commonly used as a solvent for silver chloride. Then the recommendation must take into account both silver chloride and sulfuric acid.

Body material selection. As a general rule, it is recommended that pipeline or tank material be used for the valve body whenever possible. This is particularly important when using metal screwed end valve bodies because of galvanic corrosion. However, certain throttling or control valves may require a more sophisticated material for the valve body than the pipe because of velocity or pressure drop conditions being more severe in the valves. In cases when more than one material is satisfactory for the particular service, it is usually best for the user to make a selection based on previous experience and possible variances in individual plant practices.

Other things to consider are the following:

1. Generally where cast iron is recommended for a service, ductile iron and cast steel will also be satisfactory.
2. For similar compounds such as potassium and sodium, normally the same material is suitable for either service.
3. Where abrasion-resistant materials or linings are required, a soft, resistant lining similar to a soft natural rubber should be used on abrasive mixtures of sand, silt, and/or mineral particulate matter; where abrasive fluids contain sharp, jagged particles such as iron filings or glass, it may be more desirable to use hard, resistant materials including stainless steel, ceramic, iron, or tough linings such as Tefzel or PVDF. Plastics are not generally recommended for abrasive services.

It is important to remember that not all valves are available with all materials, and depending upon the service a particular material may or may not perform as expected. Where a valve is to be used in abrasive service, its material of construction may be selected based on the results of a sand slurry test. The best method in material selection is to have a complete fluid and service profile and to use past experience to anticipate performance. Consulting a valve manufacturer is also a good practice.

Sand slurry tests are used to evaluate the abrasion resistance of materials. A rectangular sample of each material is rotated in a 50-50 sand-water slurry for 7 hours (h) at 1750 revolutions per minute (r/min). Weight loss for each material is rated relative to the weight loss of carbon steel, rated at 100. The lower the figure, the better the abrasion resistance.

Material	Weight loss*
UHMWPE	15
Nylon 6-6	31
Polyurethane (D-70)	37
TFE	72
304 Stainless steel	84
Carbon steel	100
Polypropylene	190
Polysulfone	300

*Data courtesy Hercules, Inc.

Standard materials (summary)

Ball, butterfly, and plug valves are available in a variety of material choices for bodies and sealing elements.

Body*	Seal
	Ball Valves
Alloy 20	PTFE
Aluminum	Reinforced PTFE
Brass	Bronze-filled PTFE
Bronze	Reinforced nylon
Carbon steels	Carbon graphite
Ductile iron	Kel-F
Hastelloy C	Polyethersulfone
Kynar	
Monel	Polyimide
Nickel	Metals
PVC	
Stainless steel	
Titanium	

Body*	Seal
	Butterfly Valves
Aluminum bronze	Polyethylene
Carbon steel	PTFE
Stainless steel	Reinforced PTFE
PVC	Metals
	Plug Valves
Carbon steel	Silicone based (234-dimethyl, 281-fluoro)
Cast iron	Hydrocarbon blend (147, 167, 386, 421, 654, 921)
Stainless steel	Glycol-polyol based (357, 755, 833, 852)
	Vegetable oil based (555, 555WG)
	Fluorocarbon based (660)
	Ester-polyester based (862, 950, 960)
	PTFE

*Ball, butterfly, and plug valve closure elements may also be provided in PVC and ceramic.

Chapter

2

Linear-Shaft Valves

2.1 Introduction to Linear-Shaft Valves

What is a linear-shaft valve?

A linear-shaft valve, as opposed to a rotary-shaft valve, is one that provides shutoff or control by raising or lowering the closure member into or out of the flow path. Valves considered linear-shaft valves are gate, globe, pinch, and diaphragm valves. A major feature of the linear-shaft valve is that tighter shutoff, within the limits of the materials and designs, may be achieved by tightening down on the threaded stem.

Principal linear valve types

Gate valve. This type of valve is commonly used as a stop valve to turn on and shut off the flow, as opposed to regulating flow (see Fig. 2.1). It gets its name from the gatelike disc which operates at a right angle to the flow path.

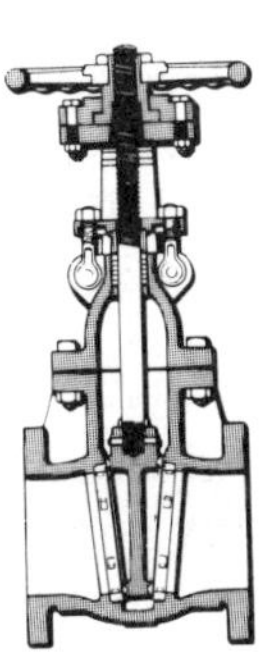

Figure 2.1 View of typical gate valve. (*Courtesy of Crane Company.*)

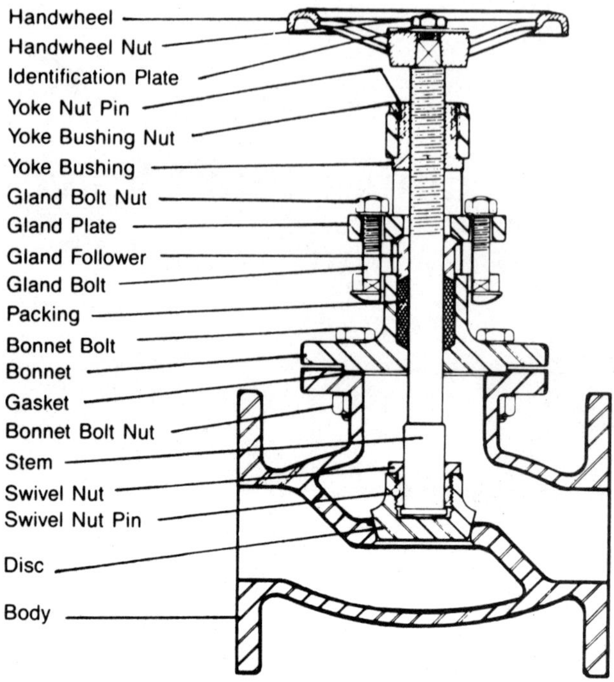

Figure 2.2 View of typical globe valve. (*Courtesy of Crane Company.*)

Globe and angle valves. The flow through a globe valve follows a tortuous path which causes increased resistance to flow and a considerable pressure drop. Because of the seating arrangement, globe valves are the most suitable for throttling flow. The valve gets its name after the globular shape of the body (see Fig. 2.2).

Angle valves are similar in principle and a companion line to the globe and are designed to permit a 90° turn in the piping system. These valves provide less of a flow restriction.

Pinch valve. A pinch valve is constructed of a section of flexible tube that provides on-off or throttling control by pinching off the flow much like the action common with a garden hose. The pinching action is provided by a pinch bar that is forced against the tube by threading a linear shaft (see Fig. 2.3).

Diaphragm valve. Like a pinch valve, a diaphragm valve controls fluid flow through the action of a linear shaft (see Fig. 2.4). Diaphragm valves are available in many different materials for a variety of corrosive and erosive applications. The shaft of a pinch or diaphragm

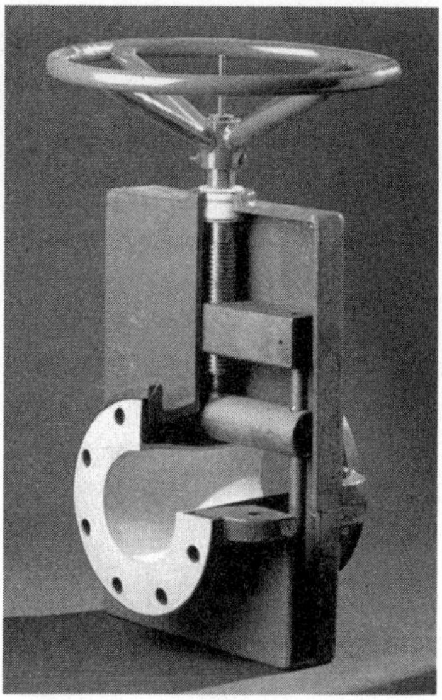

Figure 2.3 The pinch valve is constructed of a section of flexible tube. (*Courtesy of Red Valve Company, Inc.*)

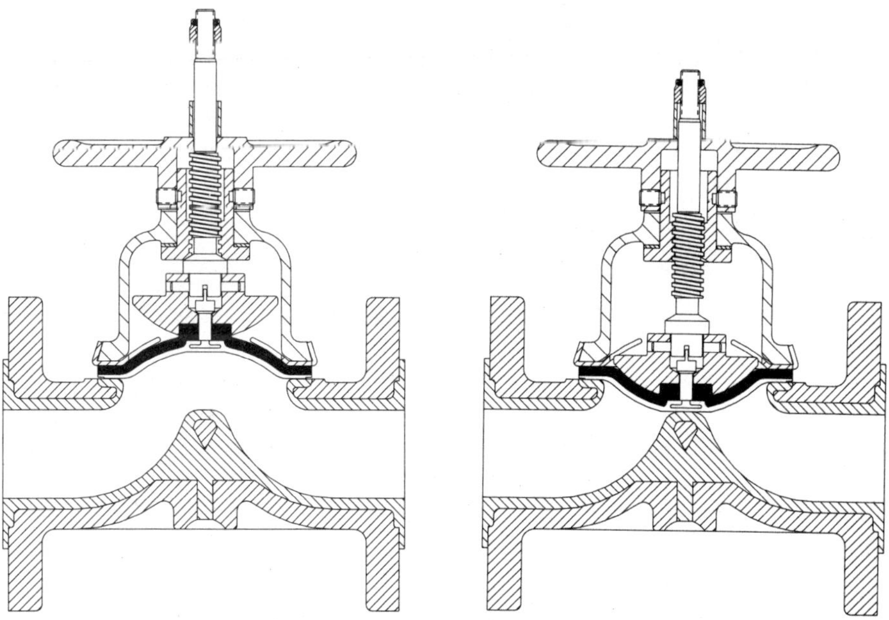

Figure 2.4 The shaft of a diaphragm valve is isolated from the flowing medium. (*Courtesy of ITT Engineered Valves.*)

valve is completely isolated from the fluid. The designs also virtually eliminate stem leakage.

Stem construction for linear valves

Depending on the style, linear valves are available with some or all of the following types of stem designs.

Rising stem—outside screw and yoke (OS&Y). This construction keeps stem threads outside the valve and is recommended where high temperatures, corrosives, and solids in the line might damage stem threads inside the valve. When the handwheel is turned, the stem rises as the yoke bushing engages the stem threads. The external threads enable easy lubrication. Care must be taken, however, to protect exposed stem threads from damage.

Rising stem—inside screw. This is the most common stem design in bronze gate valves. Since the handwheel and stem both rise, adequate clearance must be provided for operation. The stem and handwheel position indicate the position of the disc inside the valve. In an open position, the repacking seat helps protect the stem threads, but care must be taken to protect the stem externally.

Nonrising stem—inside screw. The chief advantage of this design is that it requires minimum headroom for operation. Since the stem does not travel vertically, packing wear is reduced. Heat, corrosion, erosion, and solids may damage the stem threads inside the valve and cause excessive wear. Also it is impossible to determine the disc position since the handwheel and stem do not rise.

2.2 Gate Valves

Gate valves are the most used valves in shutoff applications because of their simple design and operation. They have little obstruction to flow, low turbulence, and low pressure drop. These valves are normally used where operation is infrequent and the valve will be fully closed or fully open. They should not be used for throttling applications, because accurate control of flow is practically impossible. In addition, most of the flow change through a valve occurs near shutoff at high velocity. If the valve is opened slightly, the seat and disc are subjected to severe wire drawing and erosion that will eventually prevent tight shutoff.

Gate valves usually have a low pressure drop when fully open, provide a tight seal when fully closed, and are relatively free of contamination buildup. They are prone to vibration when in a partially open position. Response characteristics are slow, and large actuating forces are required. Gate valves should be opened slowly to prevent hydraulic shock in the line. Slow closing helps to flush trapped sediment and dirt.

Basic gate valve design

Gate valves are used widely in on-off applications. They are available in a variety of materials and are almost exclusively metal-seated valves. The disc of a gate valve is raised and lowered into and out of the flow by a threaded stem. In the open position a gate valve provides practically no resistance to flow. Because of the design they should not be used in throttling service.

The disc is connected to a threaded stem so that counterclockwise rotation of the handwheel raises the disc to open the valve and rotation in the opposite direction lowers the disc to close it. One advantage of the threaded stem operation is that the disc may be more tightly wedged into the seat if wear or trapped particles need to be overcome for shutoff. Metal-seated valves are not considered bubble-tight shutoff valves. Some leakage should be expected. Guidelines for leakage classifications are published by ANSI.

Gate valves are operated by a handwheel, gear operator, quick-action lever, or a pneumatic or electric actuator. The most commonly used operator is the handwheel which basically provides the rotating action necessary to drive the threaded stem up and down. For larger gate valves requiring higher thrust, a gear operator is used.

Gate valves used in higher-pressure applications should have a relatively loose disc-to-stem connection to eliminate any side thrust on the disc that could be transmitted to the stem and tend to strain and possibly bend it or cause premature packing wear. On gate valves the seating is perpendicular to the line of flow. That is why gate valves are impractical for throttling service and for frequent operation (see Fig. 2.5). For instance, a 6-in (150-mm) gate valve holding fluid at 300 lb/in^2 (21 bar) puts a load of over 4 tons [3600 kilograms (kg)] on one side of the disc, if there is only atmospheric pressure on the other. While seated tight, there is no wear or undue strain on the disc or seats. But each time the valve is cracked open, there is the threat of erosion of the seating surfaces by the high-velocity flow.

Gate valves are available with a variety of end connections: screwed, soldered, butt weld, socket weld, and flanged.

The most common materials used in the manufacturing of gate valves are bronze, iron, and steel. Entire valves or only critical parts can be made from more exotic and alloyed materials including glass

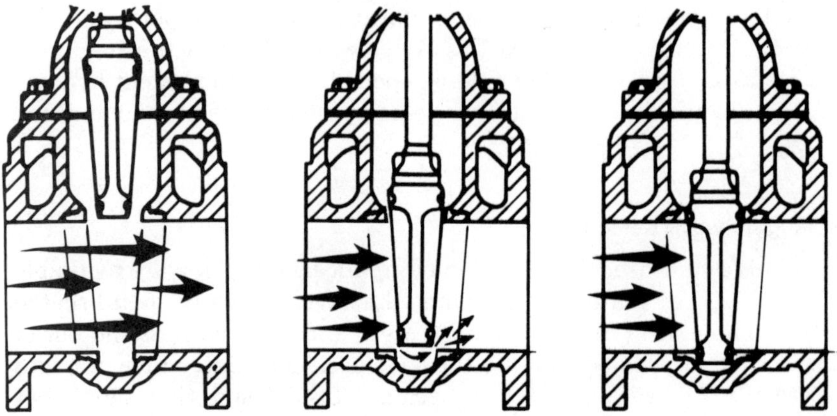

Figure 2.5 Gate valves should not be used in throttling control applications. (*Courtesy of Crane Company.*)

and plastics. The choice of materials depends on the application and the economics of construction.

Gate valves are manufactured with four disc designs that may be selected for specific applications. They are solid wedge disc, flexible wedge disc, split wedge disc, and double disc.

Solid wedge disc. Just as the name indicates, this is a solid disc valve that may be installed in any position (see Fig. 2.6). The solid wedge design is ideal for steam, water, air, oil, and gas. It is widely used in turbulent flow applications because the disc will not chatter or vibrate. It is not well suited for very high temperature applications when variations on the valve body temperature could cause the disc to jam in the seat because of uneven cooling of the body. In this case a flexible wedge disc should be considered.

Flexible wedge disc. The flexible wedge disc was developed especially to overcome sticking in high-temperature service with extreme tem-

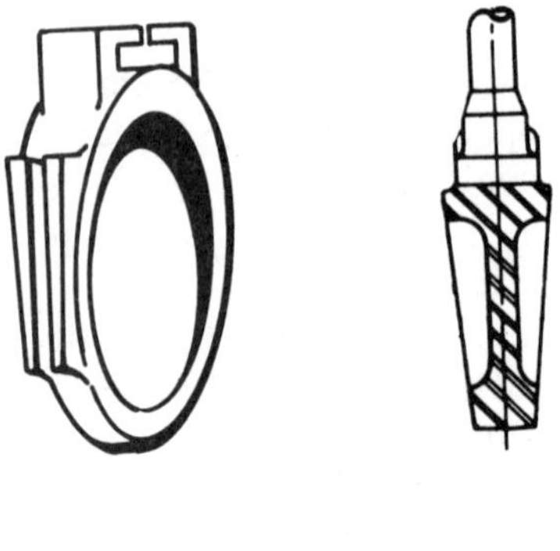

Figure 2.6 The solid wedge gate valve disc is ideal for steam, water, air, oil, and gas applications. (*Courtesy of Stockham, Inc.*)

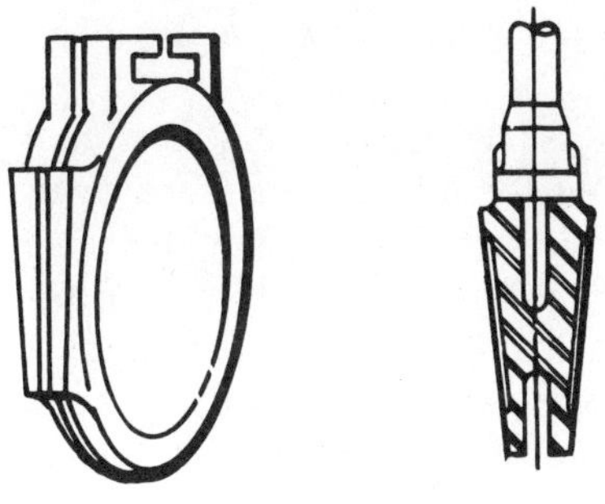

Figure 2.7 The flexible disc overcomes sticking in high-temperature service with extreme temperature changes. (*Courtesy of Stockham, Inc.*)

perature changes. The shape of the flexible disc is like two wheels on a very short axle (see Fig. 2.7). The axle, or spud, at the center of the disc is strong enough to carry the two halves of the disc together at all times and permits a degree of flexing action between them. It is this flexibility that makes the disc tight on both faces over a wide range of pressures and temperatures and assures minimum operating force. Although each disc face can move independently of the other (up to 2°) the construction is one-piece. There are no loose parts to cause harmful vibration.

Split wedge disc. This design has a two-piece wedge disc that seats between the matching tapered seats in the body (see Fig. 2.8). The two disc halves are joined in the center and fit so that the downward thrust of the stem will move them into the seat and the last turn of the handwheel will spread the discs against the seats. When operating the valve, the first turn on the handwheel relieves the force on the discs and unseats them.

Double disc. This parallel-faced double disc provides closure when positioned between matching seats in a valve body. As the valve is being closed, a lower spreader (or in some cases, a disc wedge) strikes a stop in the bottom of the body. Further closure brings the upper spreader into contact with the lower spreader so that the discs are forced out-

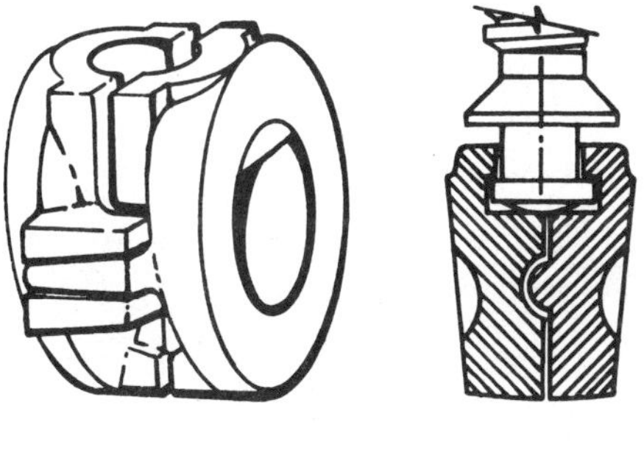

Figure 2.8 With the split wedge disc the last turn of the handwheel spreads the discs against the seat. (*Courtesy of Stockham, Inc.*)

ward against the seats. The first opening movement releases the discs, and continued operation raises them clear of seat openings.

The double disc is widely used for water service, in waterworks, and in sewage disposal plants; it is also used for oil and gas service in cross-country pipelines. It is generally unsuited for steam because rapid expansion and high velocity of steam flow tend to loosen internal parts due to vibration. This accelerates wear. If a closed valve is exposed to a rise in external temperature, this may cause a dangerous increase in internal pressure between the discs if noncompressible liquid is trapped between discs.

Because discs and body seats are perpendicular and parallel, repairing or refacing to compensate for wear is easier than on a tapered wedge disc. Double disc valves should be installed with the stem above the horizontal position. Many spreader mechanisms are subject to jamming when installed with the stem below the horizontal line.

Knife gate. A widely used version of the gate valve is the knife gate (see Fig. 2.9). This valve is used extensively in the pulp and paper industry. The thin, sharp gate is able to shear fibrous fluids. The knife gate is of lighter-duty construction than the ANSI-rated valves, is used for low-pressure shutoff, and is available in alloy or alloy-tin construction.

Gate valves (summary)

Size: 2 to 120 in (50 to 3000 mm) for on-off; ½ to 24 in (12 to 600 mm) for throttling; ½ to 6 in (12 to 150 mm) for multiorifice controlling.

Design pressure: to 3500 lb/in^2 gage (240 bar) for on-off; to 1440 lb/in^2 gage (99 bar) for multiorifice.

Design temperature: cryogenic to 500°F (530 K); −20 to 1125°F (244 to 800 K) for multiorifice type.

Capacity (C_v): for sizes ½ to 24 in, C_v = 2.5 to 8000.

Materials: most standard metals.

Connections: wafer, lugged, flanged, welded, screwed.

Variations: knife gate, V-orifice, polyurethane lined to resist abrasion, multiple-orifice for throttling, restricted trim.

Applications: on-off, solids flow, some throttling at low ΔP; multiple-orifice type for general service.

Advantages: high capacity, short length for knife gate; multiple-

Figure 2.9 The sharp edge of the knife gate valve is able to shear fibrous fluids.

orifice type for low leakage, dirty fluids, low maintenance, stability, low noise.

Disadvantages: full-port type wears rapidly in throttling service; solids are trapped under gate; packing around gate has many problems; multiple orifice type has low capacity in most sizes.

2.3 Globe and Angle Valves

Globe and angle valves are normally used where operation is frequent and flow throttling is required. They can be considered general-purpose flow control valves. A significant feature of this type of valve is efficient throttling with minimum wire drawing or disc and seat erosion. However, this type of valve is not recommended where resistance to flow and pressure drop are undesirable, because they have the highest pressure drop of commonly used valves. Their short disc travel and the few turns required for opening and closing save time and wear on the valve stem and bonnet. But the large sizes require a considerable amount of power to operate. Foreign matter on seats can usually be flushed by opening the valve slightly. In high-temperature steam service, globe valves should be installed with the steam pressure entering above the disc.

Angle valves are another form of globe valve and have similar features, such as stem, disc, and seat ring. The flow on the inlet side is at right angles to the flow on the outlet side, making a 90° change in direction. The use of an angle valve eliminates the need for an elbow and extra fittings in the line.

Basic globe and angle valve designs

Unlike the perpendicular seating in gate valves, globe valve seating is parallel to the line of flow (see Fig. 2.10). All contact between seat and disc ends when flow begins. Thus, globe valves are more efficient in throttling flow. Fully opening or closing a globe valve requires fewer turns than for a gate valve.

The stem in a globe valve not only raises and lowers the disc (or plug) but also helps to guide it into its seat. Thus, unlike a gate valve, the globe valve disc-stem connection must be relatively close fitting to prevent any extreme lateral motion of the disc that would cause it to cock and seat improperly. But, once the disc and seat are joined, the disc stops turning while seating is completed by the stem. This prevents metal-to-metal grinding between the disc and seat that can destroy the seating surfaces.

Flow-through globe and angle valves, unlike the straight-through flow of the gate valves, follows a changing course. The valves have high flow resistance and frequently create turbulence, resulting in a significant pressure drop. This characteristic makes globe and angle

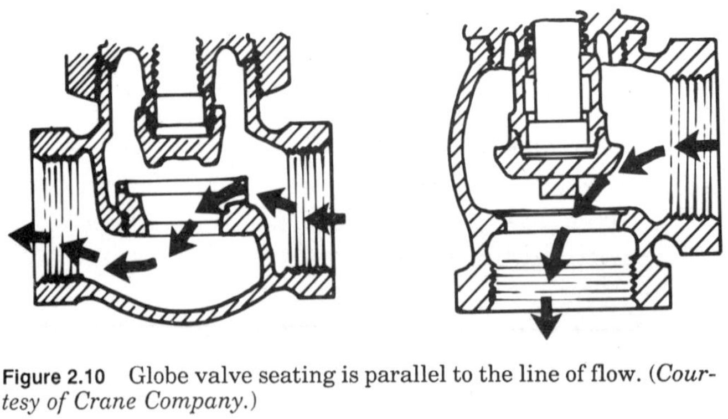

Figure 2.10 Globe valve seating is parallel to the line of flow. (*Courtesy of Crane Company.*)

valves useful for quick flow regulation that varies between tight shutoff and fully open. Globe valves are used for services requiring frequent operation or throttling service.

When manually throttling flow, the seat opening is relative to the number of turns on the handwheel. An operator can gage the rate of flow by counting the number of turns of the wheel. For instance, if it takes four turns to open fully, one turn releases about 25 percent, two turns release about 50 percent, and so on.

Two general designs are used in globe valves: the T body and the Y body. The T body is most widely used. The Y body has less change in flow direction and therefore shows less pressure drop and less turbulence, yet it still provides good throttling capability. The angle valve is a variation in design of the basic globe valve. It effectively utilizes the globe valve seating principle while providing for a 90° turn in piping. It is less resistant to flow than an in-line globe valve and because of its angle design requires fewer pipe joints, saving on installation time and labor. Screwed-in, union-flanged, seal-welded, and pressure-sealed bonnet designs are used in globe valves. The selection of the bonnet design is influenced by cost, pressure class, frequency of actuation, and installation restrictions.

Whatever wear occurs as the result of frequent or severe operation of a globe presents less of a maintenance problem than in gate valves because the seat and disc in most designs can be repaired without removing the valve from the pipeline.

Globe valve seating designs

The closure member in a globe valve is a disc moved into contact with the seat to block the flow passage between inlet and outlet. A number of disc designs are available.

In the composition disc design (see Fig. 2.11), the sealing surface is usually a flat face. Most often the disc or seat or both use a replaceable insert ring of a reinforced elastomeric material. The limits of temperature and fluid applications are restricted to the limits of the material used for the insert. The composition disc construction is good for air or gas service when tight shutoff is required but is not recommended for throttling service.

A second disc construction is all metal with a tapered or spherical surface that contacts a comparable shape in the seat (see Fig. 2.12). This design is usually used for liquid shutoff. The seal is along a line contact between the closure member and the seat. This line contact seating action normally breaks up deposits that form on the seat.

A third construction is the plug disc. There are many special configurations of plug-type discs generally used to control flow rates propor-

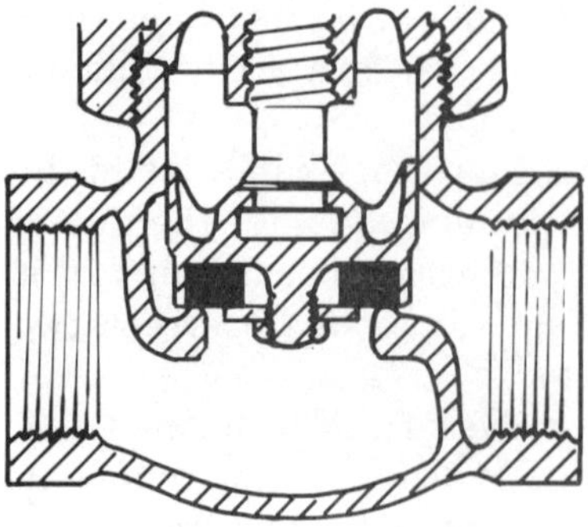

Figure 2.11 The composition disc-plug is replaceable on some globe valves. (*Courtesy of Crane Company.*)

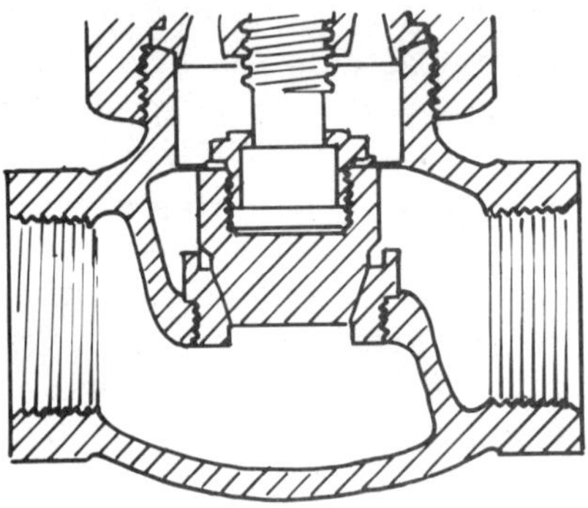

Figure 2.12 The tapered plug globe design provides metal-to-metal shutoff. (*Courtesy of Crane Company.*)

tional to the amount of stem stroke; they provide equal percentage, linear, or V-port flow control. The principal use of the plug seating arrangement in globe valves is for throttling service. The plug disc resists erosion, galling, and wire drawing; provides tight shutoff; and is excellent for high-frequency operation and throttling service.

Globe valves are made of a wide variety of materials, the most common of which are bronze, iron, and steel. Material selection is a function of the fluid to be handled, temperature and pressure in the system, and the frequency of valve operation.

When globe valves are installed, it is important, for good regulation, that the flow direction be from under the disc. The end of the valve through which both the seat and disc can be seen when the valve is closed is the under-the-disc end. The valves should be installed with the disc fully closed to prevent seat damage.

The major cause of leaks in a globe valve is the lodging of foreign matter or deposits between disc and seat. Valve leaks that occur when a globe valve is closed can often be eliminated by backing the valve off one turn and reseating, possibly repeating the action several times until the obstruction is cleared from the seat.

When a globe valve is selected, materials for body, disc, seat, and stem should be specified. The body style should also be indicated, and the operating fluid temperature and pressure determined. End connection, stem, bonnet assembly, and packing material are other variables that require careful consideration.

Globe valves (summary)

Sizes: ¼ to 16 in (6 to 400 mm); angle types to 42 in (1050 mm).

Design pressure: from 1250 lb/in^2 gage (86 bar) in large sizes to 60,000 lb/in^2 gage (4140 bar) in small sizes.

Design temperature: cryogenic to 1200°F (920 K).

Capacity (C_v): for valve sizes ½ to 16 in, C_v = 2.5 to 3600.

Materials: body and trim in most metals, smaller sizes in plastic and TFE.

Connections: screwed, flanged, welded, through-bolted.

Variations: single- and double-ported, balanced, split body, angle, Y-type, three-way, and a greater number of special forms than any other valve type.

Applications: general and extreme services, high temperature, high pressure, flashing, cryogenic, corrosive, gummy or coking fluids.

Advantages: greatest variety of types, special construction features available, large number of options, superior noise and cavitation control forms.

Disadvantages: high cost; although various flow characteristics are available, the rangeability of characterized globe valves is only moderate.

2.4 Diaphragm Valves

Diaphragm valves are used for on-off and throttling service. There are two body types: the weir and straight through (see Fig. 2.13). The weir type throttles flow in a limited range. Its throttling characteristics are those of a quick-opening valve because of the large shutoff area along the seat.

The major components of the diaphragm valve are the body, diaphragm, and stem and bonnet assembly. There is no valve seat, because the body itself acts as the seat. Diaphragm valves are frequently

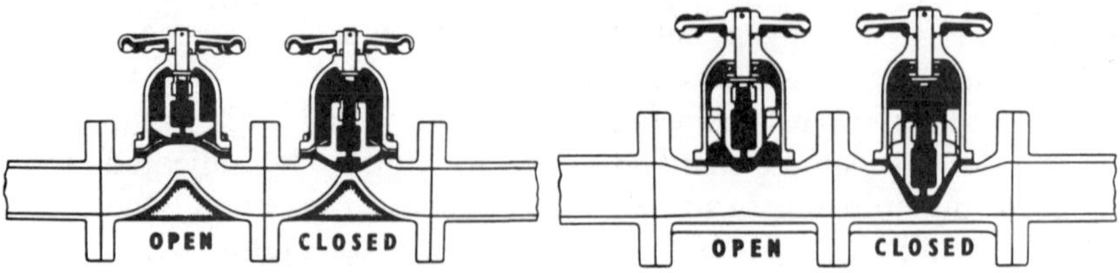

Figure 2.13 Diaphragm valves are available in weir (left) and straight-through (right) designs. (*Courtesy of Fisher Controls International, Inc.*)

used in services with highly corrosive atmospheres. Bonnets and handwheels are available with protective coatings, corrosion-resistant materials, and special lubricants.

Diaphragm valves are considered a multiturn or linear-shaft valve. They utilize a flexible element that is pressed against a fixed seating surface to block fluid flow. Diaphragm valves differ from gate and globe valves in that successful operation depends on the use of an elastomeric element to effect closure. Most diaphragm valves are available in both the straight-through and weir designs.

The operation of a diaphragm valve is relatively simple: A weir or dam built into the center of the body establishes a sealing seat for the diaphragm. The diaphragm is mechanically deflected by the action of the stem into contact with the weir.

If an application requires a valve to remain closed most of the time and to be opened infrequently, a molded-closed diaphragm can be used to reverse the bubble or hat in the center of the diaphragm. A threaded stud, molded in the diaphragm, is attached to the stem and used to raise the diaphragm from the seat to open the valve.

The advantage of using either the molded-open or molded-closed diaphragm is to increase operating life by having the diaphragm remain in the relaxed position most of the time. The diaphragm is usually an elastomeric material and may often be a composition of layers of different materials, including fabrics and metallic mesh, depending on the strength required and the process application.

Desirable characteristics of the diaphragm valve are the simple design and easy accessibility to the body interior. The open throat is easily accessible to plating or lining the body. Available lining materials include

Buna	PVC
Butyl	PVDF
Hypalon	Rubber, soft or hard
Kynar	Saran

Neoprene	Tefzel
Polypropylene	TFE

They are also available in all-plastic or all-metal construction that includes PVC, CPVC, PVDF, and polypropylene, and iron, steel, stainless steel, bronze, alloy 20, Hastelloy, Monel, and aluminum.

An important advantage of the diaphragm valve is that the stem and other components used to operate it are not exposed to the flowing medium but are totally sealed from the fluid stream by the diaphragm. The diaphragm valve is excellent for handling slurries and for low-pressure service. It is one of only a few types of valves considered by many to be leakless.

Diaphragm valves are particularly suited for handling corrosive fluids, sticky and/or viscous materials, fibrous slurries, sludges, foods, and other products that require high purity and must remain free from contamination. Many fluids that would clog, corrode, or foul the working parts of other valves pass through a diaphragm valve without causing problems. There are no packing glands to maintain and no possibility of stem leakage as long as the diaphragm does not rupture.

Hundreds of combinations of body material, linings, end connections, and bonnet assemblies combined with diaphragm selection and modes of operation add up to thousands of possible combinations for many process applications.

Diaphragm valves (summary)

Size: ¼ to 20 in (6 to 500 mm) (weir type).

Design pressure: vacuum to 230 lb/in^2 (16 bar) depending upon style and materials.

Operating temperature: −30 to 350°F (240 to 450 K).

Capacity (C_v): for sizes ½ to 14 in (6 to 350 mm), C_v = 90 to 10,000.

Materials: standard metals including iron, Monel, Hastelloy C, aluminum, and titanium; solid plastics, plastic- and glass-lined bodies; diaphragms of several elastomers and TFE.

Connections: flanged, screwed, and welded.

Variations: weir type, dual range, straight-through.

Applications: corrosives, slurries.

Advantages: simple construction, wide choice of corrosion-resistant materials, low cost, low leakage.

Disadvantages: pressure and temperature limitations, poor flow characteristics, limited rangeability, high diaphragm wear rate worsens with velocity.

2.5 Pinch Valves

Pinch valves are the simplest in design of any valve and are relatively inexpensive. They are the industrial version of the pinch cock used in the laboratory to control the flow of fluids through rubber tubing (see Fig. 2.14). These valves are ideally suited for handling slurries, liquids with large amounts of suspended solids, and systems that convey solids pneumatically. Because the operating mechanism is completely isolated from the fluid, these valves can be used where corrosion or contamination of fluid is a problem. Pinch valves are suitable for on-off and throttling service. The effective throttling range is between 10 and 95 percent of the rated flow capacity. The simple design of these valves results in a low pressure drop. The molded bodies are reinforced with fabric to improve their temperature and pressure ratings. Special valves have been designed and used with temperatures that range from −100 to 500°F (−73 to 260°C) and pressures to 300 lb/in^2 gage (21 bar).

Jacket pinch valves are used extensively for slurry control in many industries, including wastewater, sewage treatment, pulp and paper, and mining operations. The valves can handle raw sewage, sludge, slime, scum, grit, mine tailings, and other unsavory fluids. A jacket-actuator pinch valve controls flow by squeezing it, in almost the same

Figure 2.14 Pinch valves are suitable for on-off and throttling service. (*Courtesy of Red Valve Company, Inc.*)

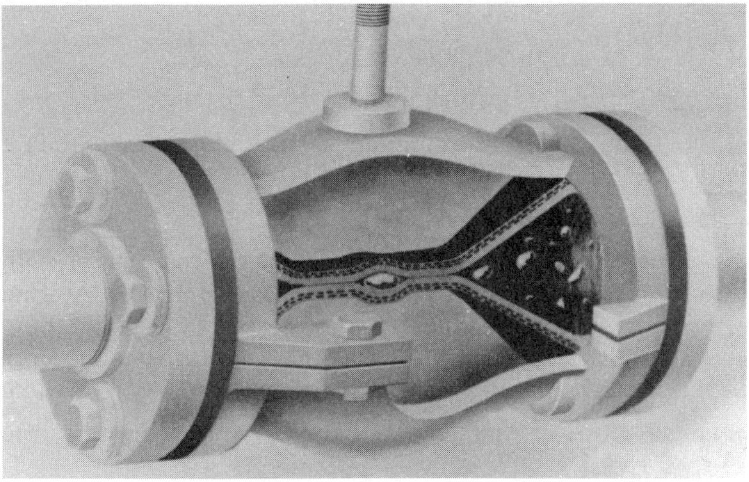

Figure 2.15 A jacket-actuator pinch valve controls flow by increasing air pressure in a jacket that surrounds the sleeve. (*Courtesy of Red Valve Company, Inc.*)

way you would control flow by squeezing a soda straw. The squeezing is accomplished by increasing air pressure in a jacket that surrounds a rubber sleeve (see Fig. 2.15). This type of valve construction works well in municipal or industrial slurries.

Since a pinch valve has no internal edges or constrictions, it does not collect sediment or debris that can cause a valve to seize or clog. Debris such as plastic bags, string, rags, or tree roots can pass directly through the valve. The flexible sleeve can seal on debris when necessary; for example, a 12-in (300-mm) jacket pinch valve can seal on a wooden two-by-four.

Pinch valve sleeves will wear out and require replacement, but the valve design is so simple that a replacement sleeve can easily be installed. First, remove the valve from the line. Next, remove the housing bolts from each side of the flange. Take out the middle housing, bolt in a new sleeve, and reverse the process. There are no alignment, anchoring, or critical fitting problems.

Although automobile tires can be repaired for additional service, a pinch valve sleeve cannot. The extreme amount of flexing, working, and high abrasion of the flowing fluid will quickly destroy any service repair. A new sleeve is the quickest and easiest maintenance method. Pinch valves are available for pipe sizes up to 30 in (760 mm).

Pinch valves (summary)

Size: ⅛ to 30 in (3 to 750 mm).

Design pressure: to 150 lb/in^2 gage (20 bar).

Design temperature: to 350°F (175°C).

Capacity (C_v): for sizes 1 to 12 in, C_v = 10 to 3800.

Materials: iron, steel, and aluminum bodies; elastomer and TFE sleeves.

Connections: flanged.

Variations: sleeve is compressed mechanically from one or both sides; clamp type is attached to sleeve to overcome low internal pressure and avoid flutter; sleeve is compressed by external fluid pressure (pneumatic or hydraulic); clamp valves have teardrop wedges placed internally to avoid crimping sleeve at closed position.

Applications: thick slurries, large particles in suspension, food processes.

Advantages: smooth sleeves can resist abrasion and corrosion. No dead pockets or crevices; straight-through path will pass large objects.

Disadvantages: poor throttling characteristics, require extremely high force to close, sleeve wears rapidly, corner folds tend to be overstressed; fluid-compressed types are limited to inordinately small differential pressures to avoid flutter; heavy-walled sleeves will accept greater pressure differentials but require still greater external pressure to close; valve C_v changes with static pressure in the line.

Chapter 3

Introduction to Rotary-Shaft (Part-Turn) Valves

In general, the advantages of rotary valves is that they are simple, lightweight, easy to automate, and easy to maintain; come in a variety of materials and end connections; come in a wide range of sizes; are available in multiport configurations; are quick opening; and can be adapted to a broad range of services.

Plug, ball, and butterfly valves are the major types of rotary valves widely used in industrial plants for flow control in fluid-handling systems. Quarter-turn valves move from fully open to fully closed with a 90° rotation of the closure member and are unique in that the flow condition is indicated by the position of the control handle or actuator.

Because plug, ball, and butterfly valves often have fluorocarbon seating materials and many are fully lined with fluorocarbon, they are excellent for corrosive applications requiring bubble-tight shutoff. They are used for a broad range of services in chemical, detergent, rubber, pharmaceutical, and pulp and paper plants; foundries; electrical component and water treatment and purification systems; breweries and distilleries; food processing plants; textile mills; and petrochemical plants. This chapter deals with rotary valves used in on-off service.

3.1 Plug Valves

Plug valves are possibly the oldest type of valve in use. Discoveries in ancient ruins indicate that a form of plug valve was used in early Roman water systems. The concept is simple: A rotary cylindrical or tapered plug with an opening through it is inserted into an open body to block or allow the flow of a fluid. When the plug is rotated, it permits flow through the opening and the fluid conduit; a quarter turn in either direction completely blocks the flow path.

The plug valve like other rotary valves requires minimum installation space, is simple to operate, exhibits fast response, and adds relatively little internal disturbance to flow. Therefore, pressure drop across the valve is low. Plug valves are not normally used for throttling, but several new designs have been used in some throttling applications. An important characteristic of the plug valve is its adaptability to multiport construction. When designed in a three-, four-, or five-way configuration, the multiport plug valve can be used to replace as many as four conventional shutoff valves. Plug valves are used throughout the industry in applications where shutoff at low pressures must be ensured and valve cavities are undesirable. There are many types and styles of plug valves used throughout the process industry: lubricated, nonlubricated, rotor, and eccentric plug.

Lubricated plug valves

Plug valves are available in either a lubricated or nonlubricated design. The lubricated valve provides tight shutoff of hard-to-handle fluids. The plug is designed with grooves, which retain a lubricant to seal and lubricate the valve and provide a hydraulic jacking force to lift the plug within the body, thereby reducing force required for rotary operation (see Fig. 3.1).

The straight passage through the port prevents accumulation of sediment or scale, and when rotated 90°, the valve plug wipes away foreign matter. The opening of the port varies between 40 and 100 percent. A full-round port opening design increases the overall plug valve body size.

Lubricated plug valves must be resealed after only a few operations—in some cases after only one cycle—and are therefore considered high-maintenance valves to be used for infrequent operation. They do not lend themselves to automation because of the uncertainty of the necessary operating torque in remote areas where maintenance is not regularly performed and because there are usually no provisions for mounting hardware.

Nonlubricated plug valves

Of the nonlubricated plug valves there are two basic designs, the sleeved and the fully lined.

Nonlubricated Teflon-sleeved plug valves. To eliminate the need to lubricate plug valves manufacturers of Teflon-sleeved valves take ad-

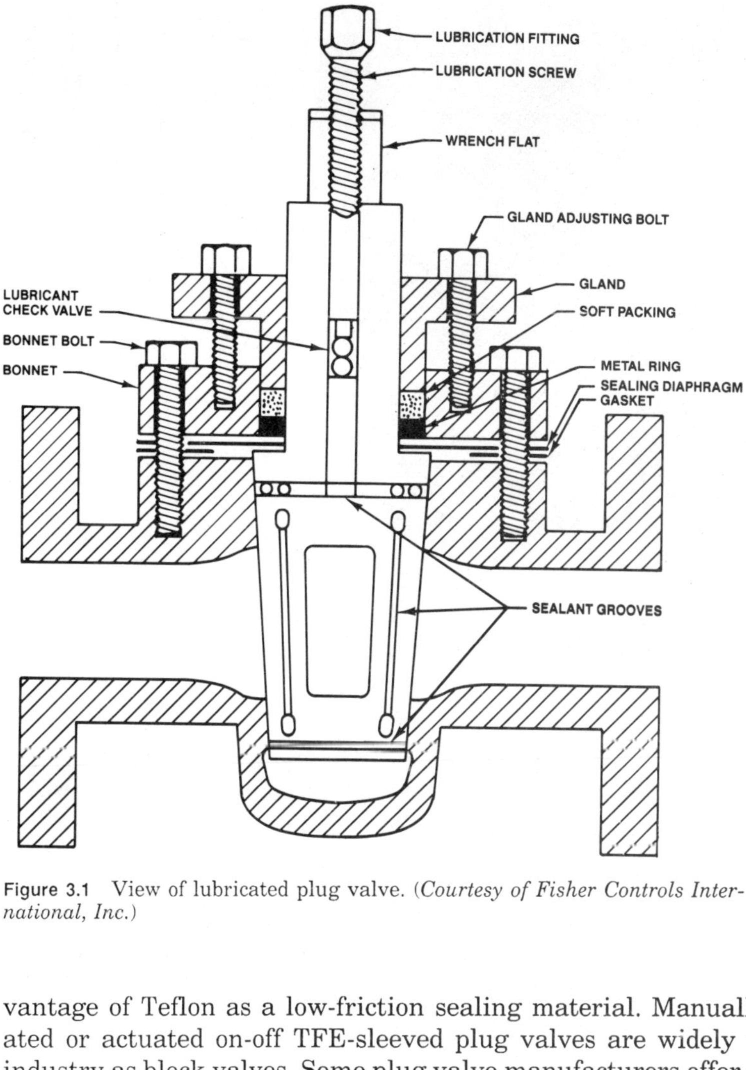

Figure 3.1 View of lubricated plug valve. (*Courtesy of Fisher Controls International, Inc.*)

vantage of Teflon as a low-friction sealing material. Manually operated or actuated on-off TFE-sleeved plug valves are widely used in industry as block valves. Some plug valve manufacturers offer special-purpose valves such as severe service valves, nuclear classes 2 and 3 valves, jacketed valves, fire-tested valves, chlorine valves, control valves, sampling valves, leak detector valves, and valves for gravity and vacuum systems.

A sleeved plug valve is one in which the plug is surrounded by a conically shaped sleeve of Teflon (see Fig. 3.2). The sleeve is first fitted into the body and acts as a seat for closure once the plug is installed. The plug is continually under pressure from the bonnet assembly to maintain sleeve and plug contact. This means that even in very low pressure applications the operating torque of a plug valve remains

Figure 3.2 Nonlubricated plug valves use a Teflon sleeve and tapered plug. (*Courtesy of The Duriron Company.*)

high. Plug valves are top-entry valves which means that the closure element (plug) and seats may be accessed through the top of the valve without removing the valve from the line. Unlike ball valves, plug valves are generally not available with a wide variety of seat materials, primarily utilizing Teflon and a few derivatives.

Some advantages of the sleeved plug valve are tight shutoff at low pressure, sleeve and seat protection at the port opening to eliminate wear, and, unlike the ball valve, there are no pockets or cavities around the plug where material may become trapped and cause the valve to seize.

Until recently the plug valve has been considered a manual valve for on-off applications. Some valve manufacturers have recently introduced characterized plugs with specially shaped port openings to provide flow control characteristics in modulating service. This option is covered more thoroughly in Chap. 4.

Materials. Plug valve bodies and closure elements can be supplied in most sizes in the following materials or combinations of materials:

Alloy 20 (CN7M)	Carbon steel (WCB)
CD4MCu	Ductile iron (A395)

Hastelloy B (N-12M)	Hastelloy C (CW-12M)
Monel (M-35)	Nickel (CZ100)
316 stainless steel (CF8M)	316L stainless steel (CF3M)
304 stainless steel (CF8)	304L stainless steel (CF3)

The TFE-sleeved tapered plug valve with the unique chemical-resistant properties of TFE and high-alloy body and bonnet construction provides the process industry with a corrosion-resistant valve, capable of tight shutoff, requiring minimal maintenance. A completely fluorocarbon lined plug valve is also available to allow the use of a lower-alloy body, bonnet, and trim materials while still providing excellent corrosion resistance.

Fully lined plug valves. The second style in which nonlubricated plug valves are available is fully lined (see Fig. 3.3). Having seen how manufacturers of plug valves, as well as many other valve manufacturers, are taking advantage of the properties of Teflon, it makes sense that it also represents an opportunity for them to use it as an alternative to high-alloy body materials. In fact, that is exactly the concept of the fully lined plug valve. A relatively inexpensive ductile iron body and

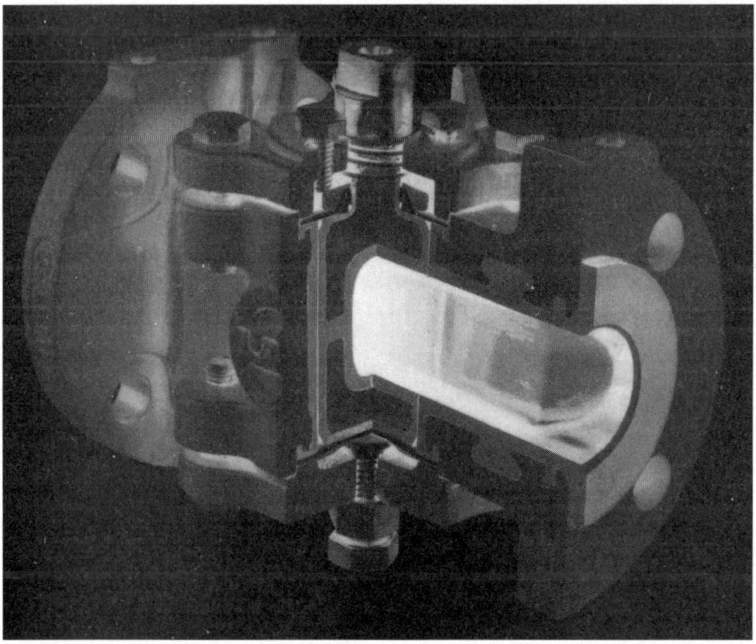

Figure 3.3 Fluorocarbon lining is locked into plug valve using dovetail design. (*Courtesy of Xomox Corporation*)

plug are fully lined and coated with Teflon. This gives the valve the ability to resist all the aggressive properties of a fluid using iron as the body material. Of course, if the external environment is equally aggressive, then an alloy or appropriately coated valve body should be used.

Some manufacturers of fully lined plug valves use dissimilar lining materials (PFA and TFE) on the sliding surfaces between the plug and liner to minimize the valve operating torque. This design utilizes the low coefficient of friction of fluorocarbon plastics to reduce the cost of an automated valve package when lower sizing torques can be used.

When selecting a fully lined valve, it is important to consider some specific service conditions to ensure valve performance. Vacuum is an especially severe service for fully lined valves. Not all manufacturers recommend their valves for vacuum service, but some may use melt processible thermoplastics that may be molded into dovetail recesses in the body to lock the liner into place (see Fig. 3.3).

If a lining is thin and is subjected to continuous abrasion, it may be destroyed, thereby exposing the less corrosion resistant metallic body to the aggressive line fluid. Some valve manufacturers use an injection molding process to vary and control lining thicknesses to suit the mechanical requirements of the various valve components.

Lined plug valve specifications

Size range: ½ to 8 in (13 to 200 mm).

Body and plug material (ASTM): A395 ductile iron lined with FEP or PFA.

Pressure rating: ANSI B16.42 class 150 (PN 20).

Temperature rating: 300°F (150°C) with FEP lining; 400°F (200°C) with PFA lining depending on the application.

Rotor plug valve

One variation on the traditional sleeved plug valve is the rotor valve (see Fig. 3.4). Rotor valves are available in ¼ through 8 in (6 to 200 mm) sizes in bronze or 316 stainless steel with a wide variety of end connections. Like the sleeved plug valve the rotor plug uses Teflon as the sealing material in a cavity-free chamber, but instead of rotating the plug within the sleeve the sleeve rotates with the plug and effectively seats against the polished internal body surface. Valves of this type are often used in sanitary applications or where an internal cavity may effect operation. Rotor plug valves are offered with many different end connections and flow patterns in the multiport design (see Fig. 3.5). Among the standard multiport pattern

Figure 3.4 Cutaway of rotor plug valve. Sleeve rotates with plug during operation. (*Courtesy of Quality Controls, Inc.*)

Figure 3.5 Multi-ported rotor plug with sanitary fittings. (*Courtesy of Quality Control, Inc.*)

selections are right angle L, three-port T, and combinations of the L and T. On a specialty basis they are available in 6- to 24-way porting patterns.

Eccentric plug valves

Eccentric plug valves are used extensively in the water and waste industry because they offer straight-through flow and high capacity

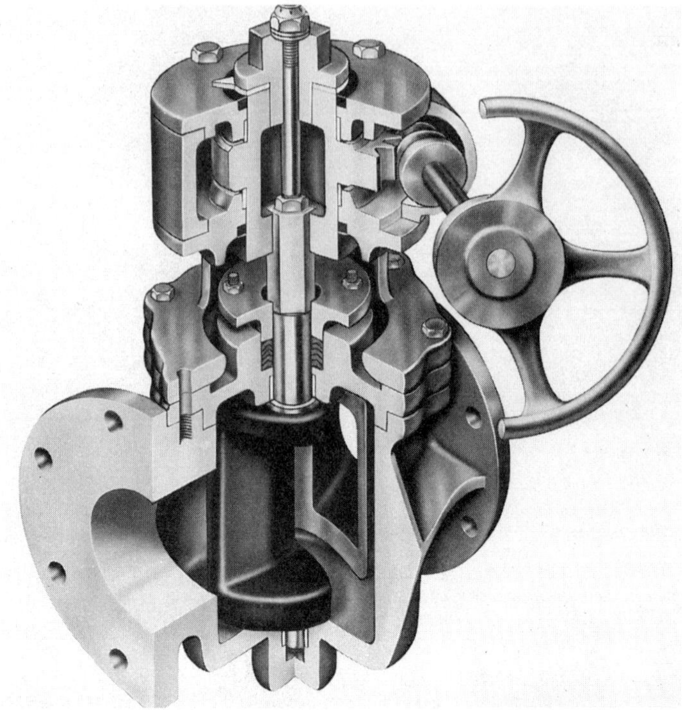

Figure 3.6 Eccentric plug valves have resilient-coated plug. (*Courtesy of DeZurik.*)

combined with tight shutoff. The design uses a resilient, coated plug segment that rotates 90° from open to closed into a raised eccentric metal body seat (see Fig. 3.6). Corrosion-resistant welded nickel, epoxy, or plastic-coated seats are used to resist corrosion. Eccentric plug valves are available with a variety of resilient plug facings to provide tight shutoff without the use of sealing lubricants, even with grit or other solid particles in the flow. As the plug closes, it moves toward the seat without scraping the seat or body walls (see Fig. 3.7). This eliminates plug binding and wear. They are available in sizes ½ to 72 in (12 to 1800 mm) and are used for the shutoff and throttling of raw sewage, grit slurries, air, water, gases, and all types of sludge.

To meet the requirements of a broad range of applications, eccentric plug valves are available with a wide selection of body materials such as Ni-resist, ductile iron, carbon steel, and stainless steel. They are available in screwed, flanged, mechanical joint, bell, or grooved end styles. Manual and powered actuators are available for shutoff and throttling control. The actuator may be a lever, handwheel, chainwheel, pneumatic or electric motor.

Eccentric Action

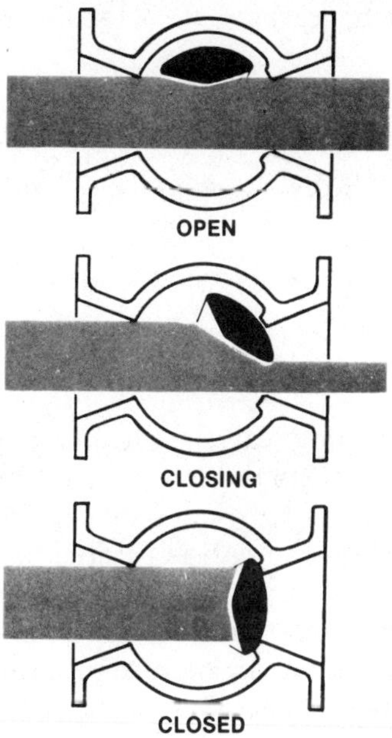

As the eccentric plug rotates 90 degrees from open to closed, it moves into a raised eccentric seat.

In the open positon, the segmented plug is out of the flow path. Flow is straight through, flow capacity is high.

As the plug closes, it moves toward the seat without scraping the seat or body walls so there is no plug binding or wear. Flow is still straight through making the throttling characteristic of this valve ideal for gasses, liquids and slurries.

In the closed position, the plug makes contact with the seat. When furnished with a resilient facing, the plug is pressed firmly into the seat for dead-tight shutoff. Eccentric plug and seat design assure lasting shutoff because the plug continues to be pressed against the seat until firm contact is made.

Figure 3.7 As eccentric plug rotates, it moves into raised eccentric seat. (*Courtesy of DeZurik.*)

Product summary (eccentric plug valve)

Sizes: ½ to 8 in (12 to 1800 mm).

End connections: Threaded, flanged, mechanical joint, grooved end.

Body materials: Ni-resist, ductile iron, carbon steel, stainless steel, and other alloys.

Pressure range: to ANSI class 600 (PN 100) depending upon style and materials.

Temperature range: −325 to 800°F (−198 to 426°C) depending upon style and materials.

Capacity (C_v): for sizes ½ to 8 in (C_v = 9 to 810).

3.2 Ball Valves

Ball valves are basically an adaptation of plug valves. Instead of a plug, a ball with a hole through one axis connects the inlet and outlet ports in the body (see Fig. 3.8). The ball rotates between seats. In the open position, flow is straight through. When the ball is rotated 90°, flow is blocked.

In addition to quick, quarter-turn, on-off operation, ball valves are compact, easy to maintain, require no lubrication, and give tight sealing with low torque. They are recommended for on-off service, mini-

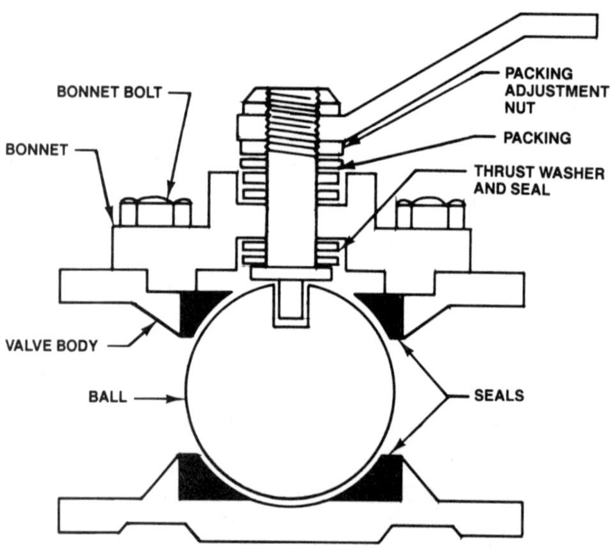

Figure 3.8 View of basic ball valve. (*Courtesy of Fisher Controls International, Inc.*)

PORT SIZES FOR LESS THAN
FULL PORT BALL VALVES

Valve Size NPS	COLUMN A Regular Port Size Inches (mm)	COLUMN B Reduced Port Size Inches (mm)
1/2	0.31 (7.9)	0.31 (7.9)
3/4	0.49 (12.4)	0.49 (12.4)
1	0.75 (19.0)	0.62 (15.7)
1 1/4	0.93 (23.6)	0.80 (20.3)
1 1/2	1.12 (28.4)	0.97 (24.6)
2	1.50 (38.1)	1.18 (30.0)
3	2.25 (57.1)	1.80 (45.7)
4	3.00 (76.2)	2.50 (63.5)
6	4.00 (101.6)	3.00 (76.2)
8	6.00 (152.4)	4.00 (101.6)
10	7.37 (187.2)	6.00 (152.4)
12	9.00 (228.6)	8.00 (203.2)
14	10.50 (266.7)	8.62 (218.9)
16	12.00 (304.8)	10.00 (254.0)
18	13.25 (336.6)	11.25 (285.7)
20	15.25 (387.3)	12.50 (317.5)
22	17.25 (438.1)	14.00 (355.6)
24	19.25 (488.9)	15.25 (387.3)

For valves above 24 NPS up to 36 NPS, port size shall be as agreed upon between purchaser and manufacturer

Tolerance = -0.06 (1.52 mm) for 12 NPS and smaller.

= -0.12 (3.05 mm) for 14 NPS and larger.

Oversize tolerances not specified.

Figure 3.9 Reduced and regular port-opening dimensions for ball valves.

mum resistance to flow, and moderate temperature service where low maintenance is required.

Ball valves are available with a ball port size that may be defined as full, regular, or reduced. Full port generally means that the opening in the ball is the same size as the inside diameter of the pipe to which it is connected. Regular and reduced port opening dimensions are defined by the MSS as shown in Fig. 3.9.

Like some other quarter-turn valves, ball valves provide shutoff in two ways: pressure-assisted and torque-seated.

Pressure-assisted design

Floating ball valve. Floating ball valves have an independent ball and stem. The bottom of the stem has machined flats. This fits into a slot

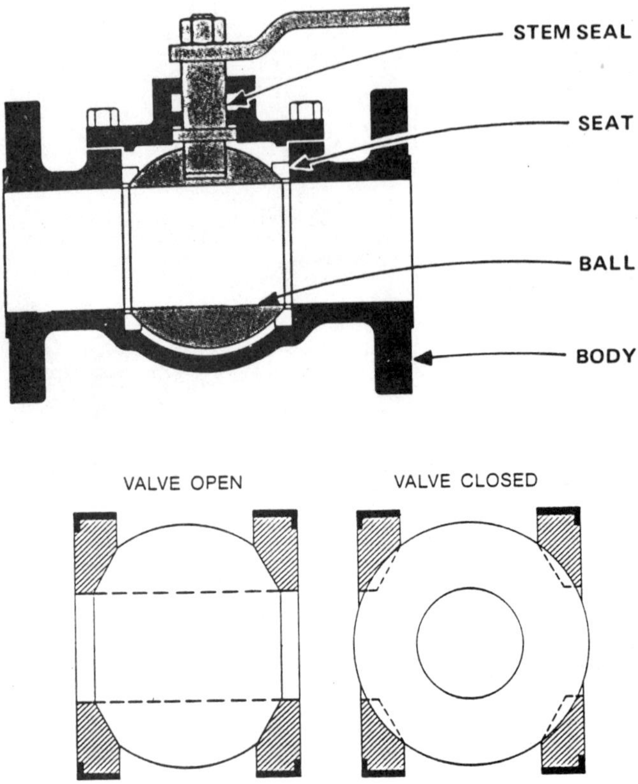

Figure 3.10 Floating ball valve is assisted by system pressure to provide shutoff.

in the top of the ball. The slot is located perpendicular to the bore through the ball so that in the closed position the ball is allowed to float downstream and press upon the downstream seat. There is some interference between the ball and seat initially which provides for low-pressure shutoff as well as to serve as support for the ball in the body cavity. As pressure increases, the shutoff capability also increases, within limits (see Fig. 3.10). While effective in providing shutoff, the loose ball and stem connection has the disadvantage of probable excessive wear and lost motion. Floating ball designs are not generally seen as a suitable choice for high cycling or when precise valve positioning is a requirement. But some options, such as a matched ball and stem, are available to expand the use of ball valves into modulating service applications.

Trunnion ball valve. Another type of ball valve that may be considered pressure-assisted is the trunnion ball valve (see Fig. 3.11). Usually used with very high pressure or large-diameter valves, the trunnion

Figure 3.11 Trunnion ball valve is supported at top and bottom to reduce operating torque. (*Courtesy of Demco Division of Cooper Industries.*)

design supports the nonfloating ball at the top and bottom, and a spring-loaded seat acts upon it to seal at low pressure. The seat is also pressurized by the line fluid to pressure-assist the seat against the ball in higher-pressure applications. One advantage of the trunnion ball valve is that it requires less torque than most valves in high-pressure, large-diameter applications.

Torque-seated design

In the torque-seated ball valve design a ball segment, which is supported at the top and bottom by stems and bearings, is rotated into an eccentrically located metal seat (see Fig. 3.12). This design achieves shutoff by continuously applying torque to the valve stem which presses the ball segment against the matching seat. Various styles of this design are ideal for throttling control, high-temperature service, or as an alternative to full-ported, high-flow, floating ball valves.

Basic ball valve styles

Ball valves are available in a few basic styles of construction. They are end entry, split body, three-piece, and top entry.

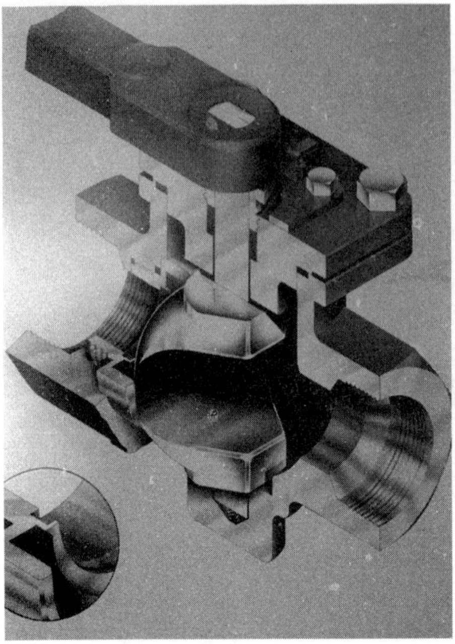

Figure 3.12 This design achieves shutoff by continuously applying torque to valve stem. (*Courtesy of Rockford Controls, Inc.*)

End-entry type. This design generally costs less than in-line maintenance valves because all internal parts are loaded through one end of the valve (see Fig. 3.13). End-entry valves are not considered to be in-line maintenance valves because the port end of the valve must be ac-

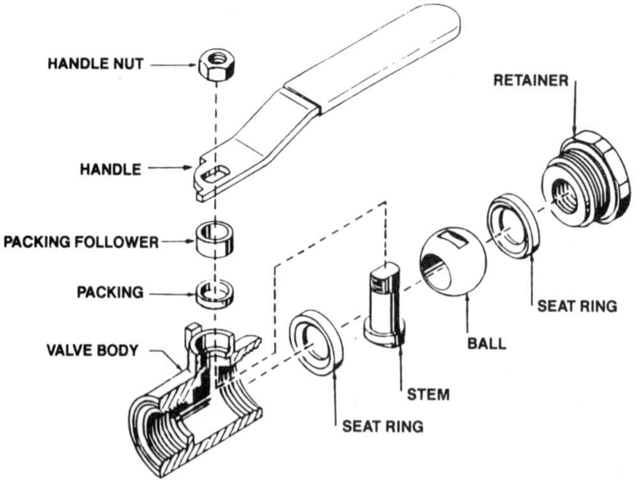

Figure 3.13 Exploded view of end-entry ball valve. (*Courtesy of Crane Company.*)

cessible in order to remove the internal parts. Some advantages of end-entry ball valves are that the one-piece body design eliminates joints and seals which are potential leak paths and the stem is internally loaded and retained and is therefore blowout proof. Seats of reinforced Teflon are securely retained in the body on either side of the ball to assure tight sealing and smooth operation. The stuffing box consists of a packing nut, spring washers, and a gland to maintain a sustained pressure on the packing. This assures a tight stem seal. Most ball valves of end-entry design are available in sizes ¼ through 2 in (6 to 50 mm). Bronze threaded-end valves are rated at 125 lb/in^2 steam (9 bar) and 500 lb/in^2 WOG (water, oil, gas nonshock) (35 bar). Carbon and stainless-steel threaded-end valves are good for 125 lb/in^2 steam (9 bar) and 800 lb/in^2 WOG (55 bar). Flanged valves are available in classes 150 and 300 (PN20 and PN50) in carbon and stainless steel. End-entry-type ball valves are not available in a full-port design.

Split-body ball valve. Ball valves, particularly full-port, flanged valves, are available in a flanged, split-body design (see Fig. 3.14). They are designed for class 150, 300 (PN20 and PN50), and higher, threaded or welded pipe ends, full-port or regular port, typically in a size range of ⅜ to 12 in (10 to 300 mm) in stainless or carbon steel. Design features include integral seat, blowout-proof stem, and in some cases a fire-safe design.

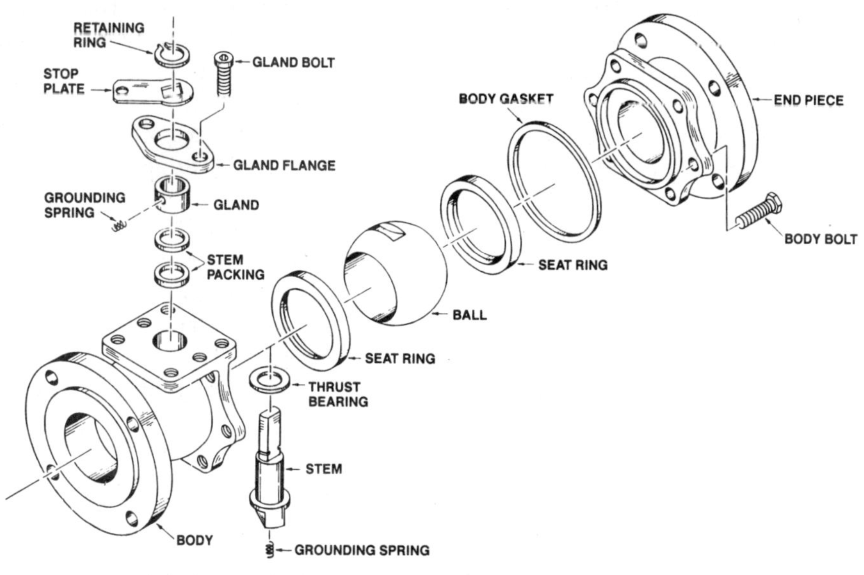

Figure 3.14 Exploded view of split-body ball valve. (*Courtesy of Crane Company.*)

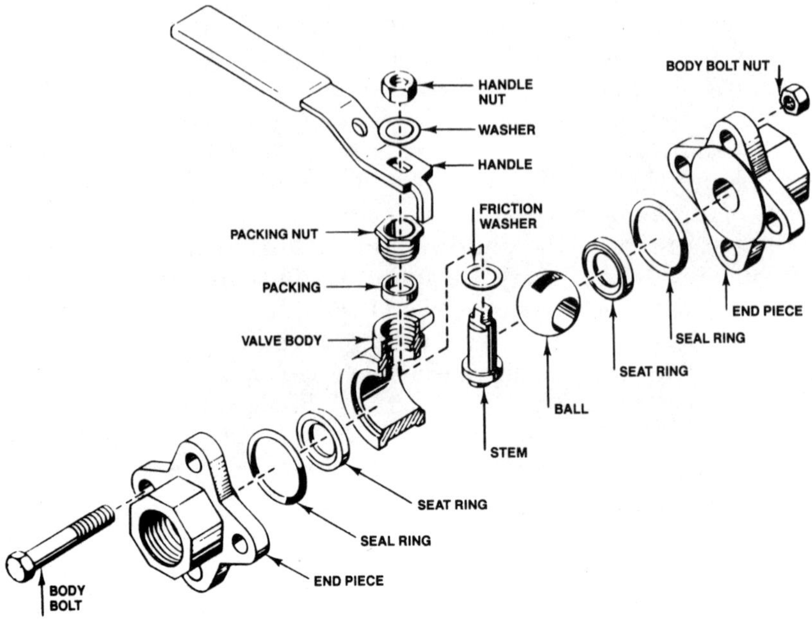

Figure 3.15 Exploded view of three-piece ball valve. (*Courtesy of Crane Company.*)

Three-piece type. The three-piece ball valve consists of a center section (the body) that contains the ball, stem, and seals, and two ends that are bolted to the center section using body bolts (see Fig. 3.15). The three-piece ball valve features in-line maintenance—the center may be removed and repaired without disturbing the pipe connections or cutting the line. These valves are widely available in carbon steel, carbon with stainless-steel trim, and stainless steel in ¼ to 2 in (6 to 50 mm) sizes. Larger sizes are available. All sizes are offered with a wide variety of seat materials, and most are readily adaptable for automation.

In addition to in-line maintenance, three-piece ball valves may be supplied with all types of end connections. These may include screwed ends, butt weld, sanitary ends, socket weld, and flanges. These end connections may be supplied in identical pairs, with the same connections on each end of the valve, or in unlike pairs. This means that one end may have, for instance, a flanged connection with a butt-weld connection on the other.

Regarding in-line maintenance, care must always be taken when loosening any bolts on any valve installed in a line to be sure that fluid under pressure has been first relieved.

Top-entry ball valve. Another in-line maintenance ball valve is the top-entry valve (see Fig. 3.16). This design has a one-piece end-to-end

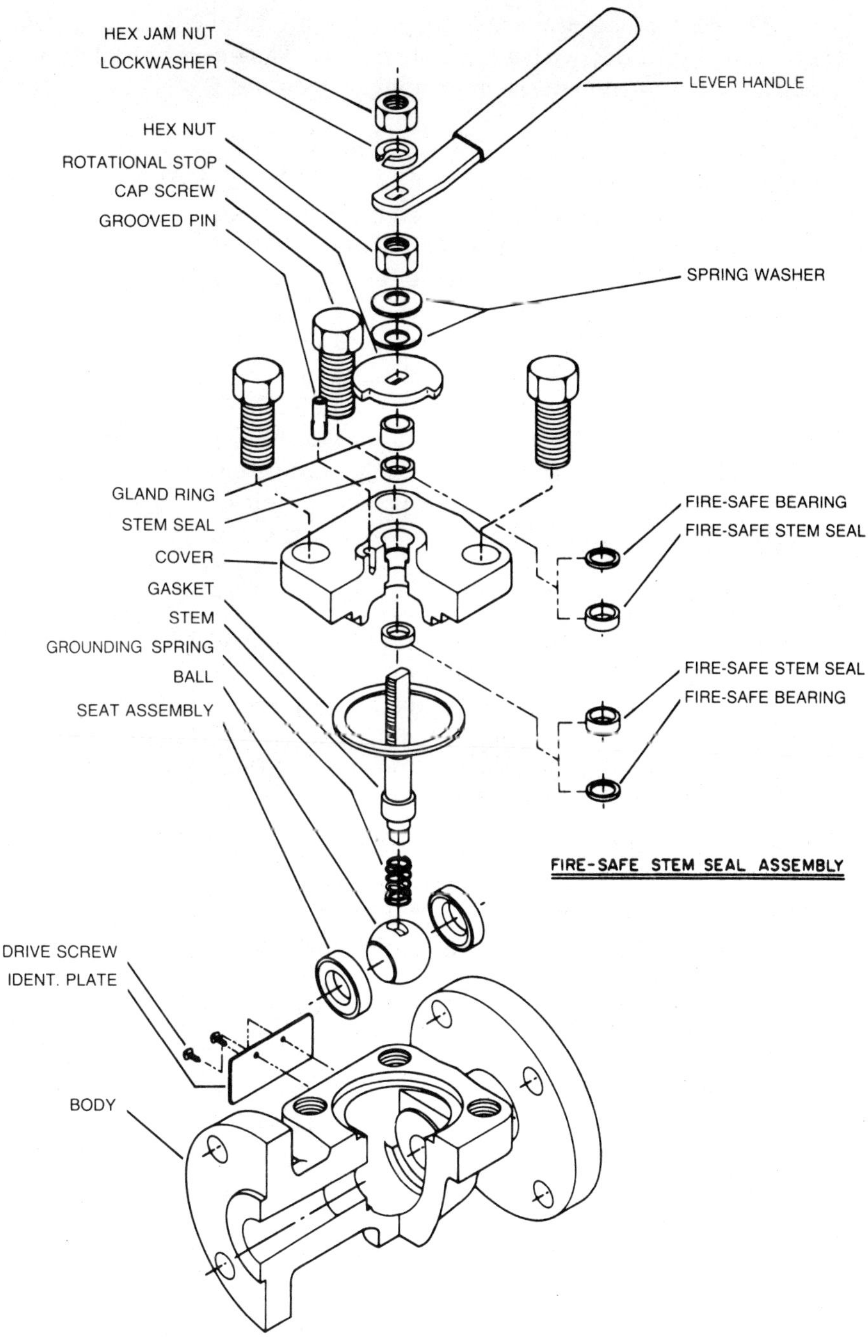

Figure 3.16 Exploded view of top-entry ball valve. (*Courtesy of ITT Engineered Valves.*)

body. The ball, stem, and seals are loaded into the body through the top and then retained with a top body bonnet. Top-entry valves ensure that the pipe alignment is maintained when service is being performed.

Ball valves are available in many different materials of construction as outlined below. This has led to the ball valve's extensive use throughout many industries. In addition to alloy materials, ball valves may be fully lined or made entirely of PVC, CPVC, or other plastics. Their performance is further enhanced by the many seat materials that are used with ball valves.

Ball valves are available in multiport configurations, similar to plug valves. These valve bodies are usually larger than the equivalent straight-way valve, because a larger-diameter ball must be used to permit multiport drillings. In addition to the handle position being used as an indicator of the valve setting, on multiported valves there are grooves cut into the top face of the stem to show the type of diverting element used and the flow path through the ball.

Ball valves (summary)

Sizes and ratings: floating ball from ½ to 36 in (12 to 900 mm) for class 150 (PN20), or 12 in (300 mm) for class 2500 (PN 400); segmented ball from 1 to 24 in (25 to 600 mm) for class 150 (PN 20) to 16 in (400 mm) for class 300 (PN 50), and to 2 in (50 mm) for class 600 (PN 100); trunnion ball from 1½ to 72 in (40 to 1800 mm) for class 600 (PN 100).

Temperature and pressure differential limits: highly variable; depend on size and materials of construction of seals, packing, and bearings; maximum shutoff differential is 1500 lb/in^2 (100 bar) with standard actuators; maximum operating temperature is over 1400°F (760°C).

Capacity (C_v): for standard ball sizes ½ to 20 in C_v = 15 to 22,000; for full bore ball sizes ½ to 20 in C_v = 26 to 73,000; for segmented ball sizes ½ to 4 in C_v = 10 to 1343.

Materials: body and ball are iron, bronze, steel, stainless steel, alloy 20, aluminum, Monel, titanium, and other metals as specials; also reinforced plastics and glass are used; balls are TFE-faced; seals may be TFE, Stellite, or special metals.

Connections: flanged, through-bolted, screwed, welded.

Variations: segmented and characterized; standard bore (approximately 80 percent of inlet diameter); full bore; restricted trim; hollow ball; three-, four-, and five-way bodies assembled in one to three pieces; stem attachment may be loose or integral; floating ball moves against downstream seat; trunnion type is supported by ball extension pieces; various anticavitation and antinoise constructions; liners and coatings to resist erosion.

Applications: slurries, on-off control, large size, pigging services, low-pressure drop requirement.

Advantages: high capacity (low pressure drop), low leakage, moderate cost, wide rangeability, two-stage throttling, pigging, or other full-flow capability.

Disadvantages: high torque, high friction (deadband), disposed to cavitation and high noise level, vulnerable to seal wear, most designs have dead space in body cavity (at least one does not).

3.3 General-Purpose Butterfly Valves

Butterfly valves are built on the principle of a pipe damper (see Fig. 3.17). The flow control element is a disc of approximately the same diameter as the inside diameter of the adjoining pipe and rotates on a vertical or horizontal axis. When the disc is parallel to flow, the valve is fully open. When the disc is perpendicular to flow, the valve is shut. Throttling can be achieved by rotating the disc to an intermediate position and locking the handle in place or using a self-locking actuator. Many butterfly valves, the ones we shall discuss here, are produced with a resilient seat that seals against the disc (see Fig. 3.18). This arrangement provides a leak-tight closure and in most cases eliminates the need for flange gaskets.

Butterfly valves in larger sizes have weight, space, and initial cost advantages over gate, globe, plug, and ball valves. Maintenance costs

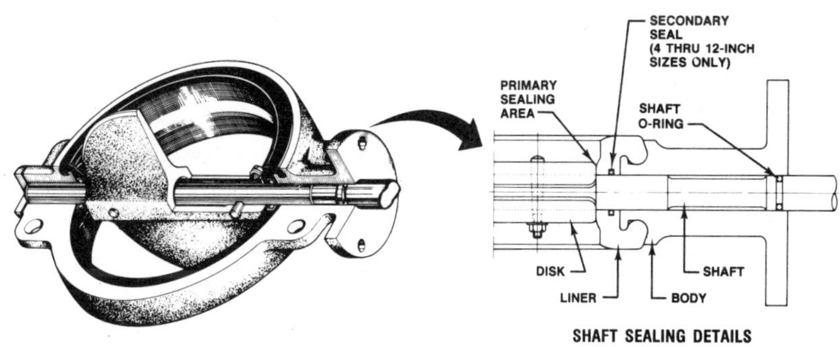

Figure 3.17 View of basic butterfly-damper valve. (*Courtesy of Fisher Controls International, Inc.*)

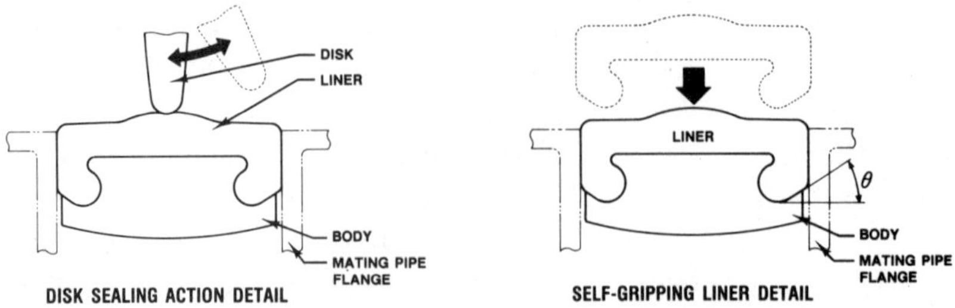

Figure 3.18 Disc-to-seat contact of butterfly valve achieves shutoff. (*Courtesy of Fisher Controls International, Inc.*)

are low because of the minimum number of moving parts and the absence of pockets that trap fluids.

They can be operated manually, mechanically, or automatically and can be used in handling a variety of media, including liquids, solids, slurries, gases, and vapor (steam).

The most economical body design is the wafer type, which fits between two pipeline flanges. Butterfly valves are available with flanged ends, for bolting to pipe flanges, and with lug body, screwed-end, victaulic and even weld-end construction. Replaceable discs are available in a wide range of materials, including most metals. Metal discs may be covered with corrosion-resistant elastomers and plastic.

Body types

General-purpose resilient-seated butterfly valves are manufactured in three basic valve body styles: wafer, lug, and two flange.

Wafer. The wafer body is designed for quick installation between pipe flanges. Resilient-seated butterfly valves require no gaskets since the liner face acts as the gasket. The body is generally self-centering inside the flange bolt circle (see Fig. 3.19).

When installing a butterfly valve, proper alignment with the pipe opening is important. When the disc is in the open position, it must clear the adjacent pipe (see Fig. 3.20).

Lug wafer. This design is basically the same as wafer except for the addition of lugs around the body (see Fig. 3.19). For end-of-line suspension, lug-type valves should be selected. Often, butterfly valves are used to isolate other equipment in the line, downstream of the valve, for periodic maintenance and repair. This application requires a lug-body valve with a blocking capability that will withstand system pressure and seal the line during the maintenance period.

Without a downstream flange or spool piece, some lug-type valves

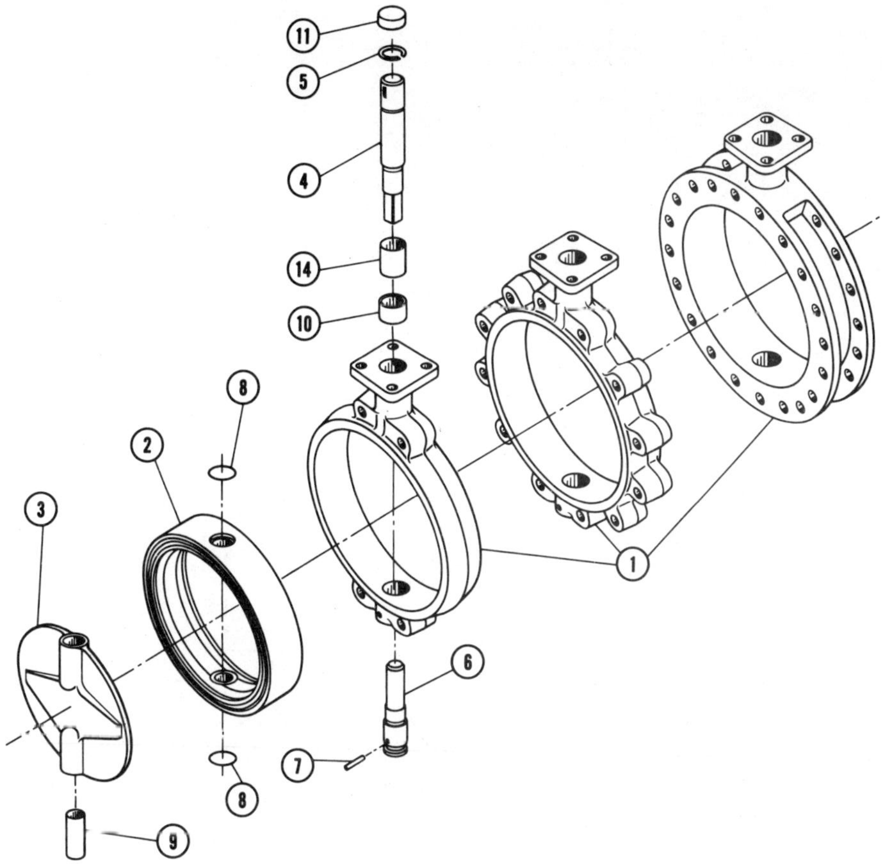

Figure 3.19 Butterfly valves are available in wafer, lug, and two-flange designs. (*Courtesy of Demco Division of Cooper Industries.*)

are derated for safety to 75 lb/in^2 (5 bar) working pressure when used for end-of-line suspension. Full valve rating may be restored by temporarily installing a downstream flange. These valves are available with the lugs either tapped or through-drilled.

Two flanged. Similar to the lug body, the two-flanged valve offers integrally cast flanged ends for installation similar to flanged gate valves. A gasket is required for resilient-seated two-flanged butterfly valves.

Seating materials

The seat and disc constitute the "heart" of a butterfly valve. The material from which these parts are made, therefore, becomes important.

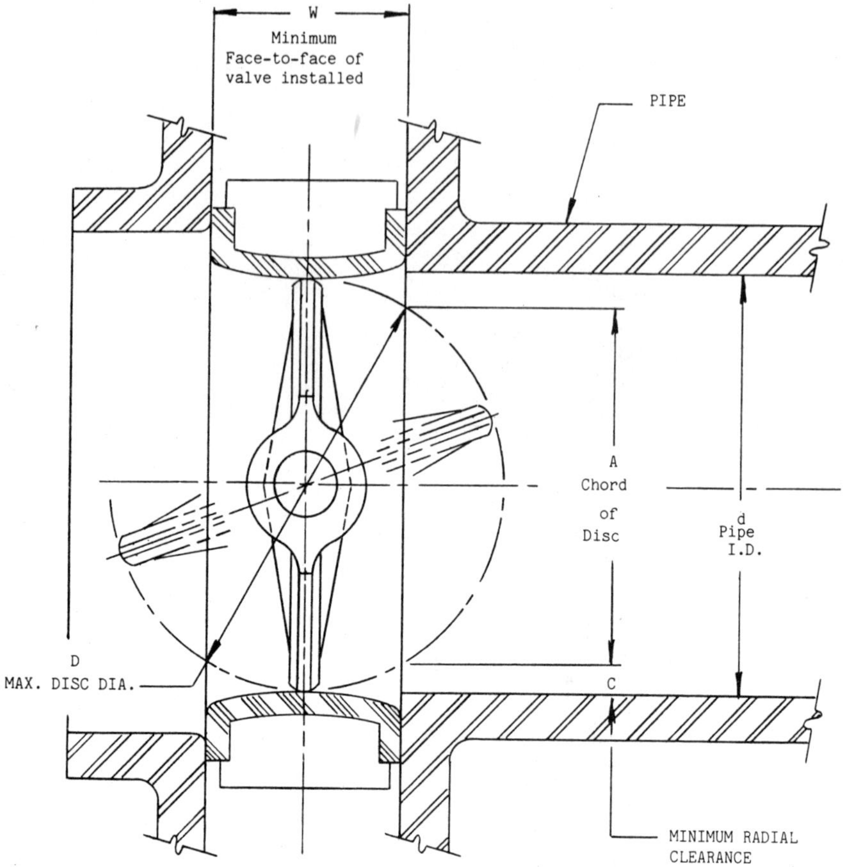

Figure 3.20 Disc-to-pipe clearance for butterfly valve. (*Courtesy of Manufacturers Standardization Society.*)

The tougher the service, the more severe the demand on seating. Valve manufacturers recognize this fact by providing a wider choice of seating materials as valves go up the pressure-temperature scale or are offered for more rigorous service.

For relatively low pressures and temperatures and for ordinary fluids, seating materials are not a particularly difficult problem. Nonmetallic composition discs are available for the tight seating needed with hard-to-hold fluids such as air or gasoline.

The selection of seating materials for corrosive fluids, regardless of pressure and temperature, is almost endless. There are many types of alloys as well as linings or coatings for use with butterfly valves. In some corrosive or erosive applications, valve components may be protected from process fluids by using a variety of elastomer liners along with coated discs. In mildly corrosive low-pressure applications, an inexpensive elastomer-lined body may be combined with a ductile iron disc or coated iron or steel disc for long life and tight shutoff. For more demanding applications, the elastomers are carefully selected for compatibility with specific process fluids, operating temperatures, and pressure drops. Liners are sometimes used to resist or slow erosion. PTFE- or PFA-coated discs are used with some lined butterfly valves to help provide increased corrosion and/or erosion resistance to a variety of process media (see Fig. 3.21).

In abrasive service a butterfly valve and disc may be coated with UHMWPE for maximum abrasion resistance plus exceptional chemical resistance. Some valves combine the UHMWPE disc and liner with a high-strength, reinforced plastic body that cuts valve weight by 70 percent or more over conventional iron body butterflies. Especially recommended for plastic piping systems, these valves are available in sizes from 2 to 12 in (50 to 300 mm) and can be used at temperatures up to 180°F (82°C). The unique nonmetallic body offers external corrosion resistance.

A variety of shaft sealing methods is used to prevent leakage of the process fluid to the environment. With most resilient-seated butterfly valves the disc-liner interference provides the primary shaft seal. A secondary seal is located between the liner and shaft. An O-ring be-

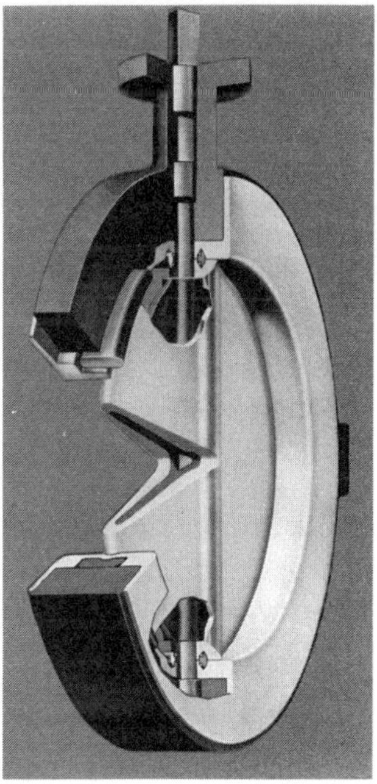

Figure 3.21 Butterfly valves may be fully lined for corrosive applications. (*Courtesy of Garlock, Inc.*)

tween the body and shaft prevents dirt or other materials from entering the environment.

Resilient-seated butterfly valves are not pressure-assisted. Shutoff is achieved by the interference fit between the disc and the resilient liner. For high-pressure applications up to the ANSI class ratings high-performance butterfly valves are often better suited.

Butterfly valves (summary)

Sizes: 2 to 36 in (50 to 900 mm) (larger sizes available in fabricated and special designs).

Design pressures: to 6000 lb/in^2 gage (400 bar).

Design temperatures: cryogenic to 2200°F (120°C) depending upon materials.

Capacity (C_v): for sizes 2 to 36 in, C_v = 130 to 70,000.

Materials: bodies available in any metal, linings, and encapsulants of most any elastomer or fluoropolymer.

Connections: wafer, lugged, flanged, and welded; for high temperatures, use lugged type instead of wafer type to reduce differential thermal expansion between body and bolts.

Variations: swing-through, lined, encapsulated, pressure sealed, low-torque discs, discs to reduce cavitation and noise, outboard bearings, and many special disc forms and seal forms.

Applications: general service, large size, viscous liquids, equal percentage flow characteristics over a wide range, throttling and on-off service, high temperature, some slurries.

Advantages: low cost, high capacity, wide rangeability (equal percentage only), self-flushing straight-through flow pattern, wide materials selection, low friction (unlined only), unidirectional torque.

Disadvantages: high torque; disposed to noise and cavitation; require more detailed specification than most valves.

3.4 High-Performance Butterfly Valves

High-performance butterfly valves include shafts, discs, and bodies that are similar in principle to the general-purpose butterfly valves previously discussed (see Fig. 3.22). Their high performance results from the incorporation of an offset (eccentric) disc in conjunction with pressure-assisted seats. This combination offers improved throttling and shutoff capabilities.

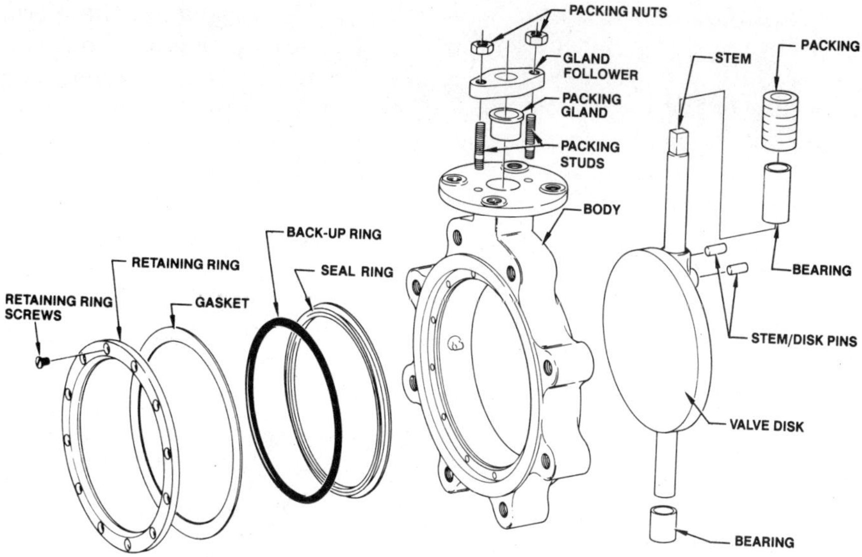

Figure 3.22 Exploded view of high-performance butterfly valve. (*Courtesy of Fisher Controls International, Inc.*)

Offset disc arrangements

Manufacturers of high-performance butterfly valves offer different offset disc arrangements. Two common styles are the single-offset and double-offset discs.

Single-offset disc. In a single-offset eccentric disc design (see Fig. 3.23) the disc face is located on one side of the rotary shaft. The shaft

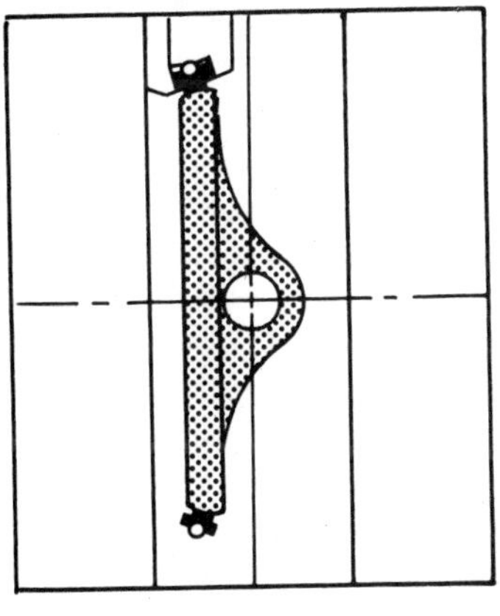

Figure 3.23 Single-offset disc.

is located along the centerline of the valve body opening; therefore, the disc face does not pull away from the seal as it is rotated. The single-offset disc design provides effective high-pressure shutoff, but because of the continuous disc-to-seat contact, this design is often specified for on-off applications only.

Double-offset disc. Double-offset discs (see Fig. 3.24) are also located on one side of the rotary shaft. The second offset occurs because the shaft is positioned to one side or the other of the valve body centerline. The valve disc moves into the seat with a "camming" action which keeps the closure contact area of the disc clear of the seat except at the final few degrees in the shutoff position. Drawing the disc away from the seat eliminates the possibility of permanent depressions in the seal often caused by prolonged disc-to-seat contact. These features reduce seal wear and promote long seal life in on-off and throttling control.

Pressure-assisted seats. High-performance butterfly valves use pressure-assisted seats to effect tight shutoff (see Fig. 3.25). These seats are generally designed so that process pressure from either direction increases the contact loads between seat and disc. The pressure-assisted seat increases its ability to resist leakage as process pressure increases.

PTFE seats are generally used for shutoff up to ANSI class VI at temperatures up to 450°F (230°C). This translates to a leakage of 1 to 45 bubbles per minute in sizes from 1 to 8 in (25 to 200 mm). This is the best shutoff specification in ANSI B16.104-1976. The seats are supported by stainless-steel springs, elastomers, or flexible plastics

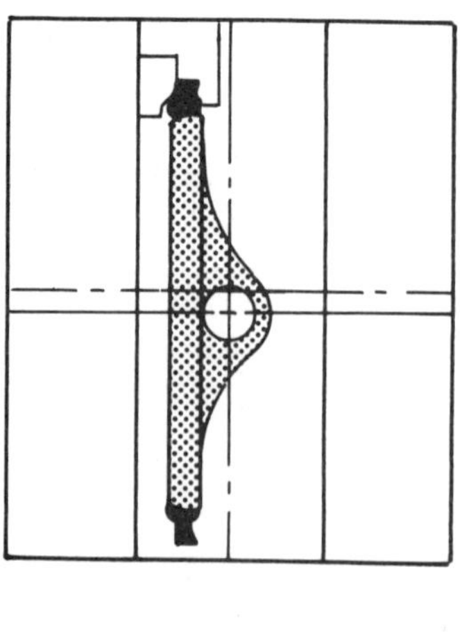

Figure 3.24 Double-offset disc.

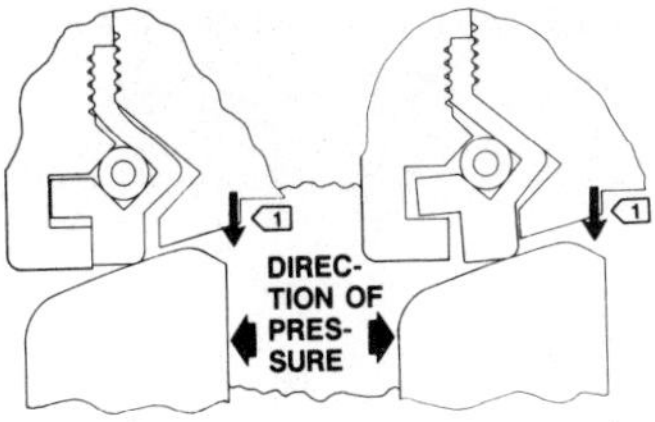

Figure 3.25 High-performance butterfly valves use system pressure for shutoff. (*Courtesy of Fisher Controls International, Inc.*)

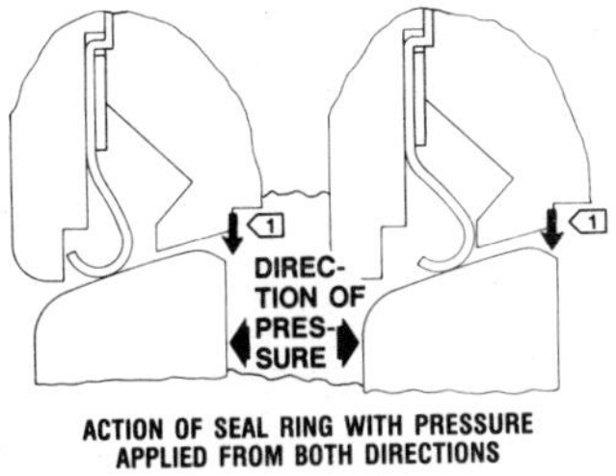

Figure 3.26 For high temperatures metal seats are used. (*Courtesy of Fisher Controls International, Inc.*)

which provide resilience to the PTFE material to compensate for wear and distortion.

Metal seats are generally specified for high-temperature service. Seats made of 316 stainless steel are effective at temperatures up to 1000°F (540°C). Metal seats are installed in conjunction with gaskets or backup rings with equivalent temperature ratings (see Fig. 3.26). Metal seats do not generally offer full ANSI class VI ratings.

High-performance butterfly valve seat materials are available for fire-rated, cryogenic, and other unique applications.

Typical applications

High-performance butterfly valves are used throughout the process industries. They are often used to replace heavy-duty conventional butterfly valves because of cost and performance advantages. When they can meet application requirements, they are often substituted for more costly globe valves, especially in sizes of 3 to 8 in (75 to 200 mm) or larger. These valves are also used in applications requiring tight shutoff which can be maintained at relatively high temperatures.

High-performance butterfly valves are available in wafer and lug-wafer designs in ANSI classes 150, 300, and 600 (PN 20, 50, 100).

High-performance butterfly valves (summary)

Note: *High performance* is a term generally used to indicate that the valve has features that enable it to perform certain tasks that are beyond the capability of standard valves. Some models operate at high pressures and with better than class VI leakage rate. Some are best suited for on-off service, while others are designed especially as control valves and to outperform standard butterfly valves in other respects also.

Sizes: 2 to 12 in (50 to 300 mm).

Design pressures: to 1480 lb/in^2 gage (102 bar).

Design temperatures: cryogenic to 1700°F (926°C).

Capacity (C_v): for sizes 2 to 12 in, C_v = 80 to 5400.

Materials: steel, alloy steel, stainless steel, alloy 20, Monel, Hastelloy C bodies; metal or TFE seals.

Connections: wafer, lugged, flanged.

Variations: offset discs of various forms, inherent linear-flow characteristics.

Applications: general service, high-pressure differential, leak-tight, large size.

Advantages: low to moderate cost, high capacity, low leakage, low unidirectional torque, low friction (offset-type discs), low seat sear, wide rangeability.

Disadvantages: propensity for cavitation and noise.

Chapter

4

Valve Designs for Special Applications

4.1 Rotary Valve Designs for Control Applications

In a process control system the valve is considered the final control element. Where once sliding stem valves such as globe or cage guided valves dominated control applications, design and materials enhancements of today's rotary valves have made them more widely accepted for such use. Heavier stems, new bearing designs, characterized control elements, higher flow capacity, and overall compactness have all contributed to this trend. Another advantage of a rotary stem control valve is the improved performance obtained because the valve stem does not move through the packing. Sliding stem movement tends to drag along some of the line fluid, creating a leak, or to react with moisture in the atmosphere, causing corrosion. In addition, improvements in actuators and positioners and the need for more valve status feedback have made rotary valve performance and economy an attractive benefit and incentive for their use. For the control valve user there are two general categories of rotary valves available for control applications: standard and characterized. For our purpose here a *standard* rotary valve is considered one that is used primarily for on-off service but that may be fitted with an actuator and positioner for modulating service. One of the benefits of using standard valves in control service is that they may also provide full shutoff if necessary. Care must be taken however to assure that the control service does not cause damage to the seat. A *characterized* valve has a specially designed closure element that may be a V-notch, segmented ball, or other such modification. The application and design of sliding stem valves in control service is well documented, so we shall limit our coverage to the above-mentioned rotary-style valves.

Process system analysis

Before a control valve is selected, a thorough analysis of the process system should be performed. To be considered are flow and fluid characteristics; system extremes, such as minimum and maximum flow disturbance and power or signal failure; toxicity, and even the piping system.

The following questions should be considered.

1. What is being controlled (pressure, temperature, pH, flow rate, etc.)?
2. What is the control range? How closely must it be controlled?
3. Is a valve the best choice for control or should a metering pump or variable-speed drive be considered? Will a simple regulator do?
4. What are the properties of the fluid (corrosive, abrasive, toxic, flammable, viscous)?
5. What are the maximum and minimum pressures (and pressure drop) and temperature? Is there a vacuum?
6. For valve sizing, what are the density and specific gravity (gas, vapor, or liquid)?
7. What are the best end connections and pressure ratings [flanged (ANSI, 150/300#, etc.), welded, threaded, wafer] for piping system compatibility?
8. Should the valve provide bubble-tight shutoff?

A very important consideration in the selection of the best control valve—based on performance and economy—is the degree of control. If it is determined that a very slight deviation from the optimum control point will result in very undesirable consequences, then it is likely that a characterized rotary or sliding stem control valve—though possibly quite expensive—is the best choice. However, as industry attempts to upgrade all aspects of process systems there are more situations where standard rotary valves can provide an acceptable degree of control as required. Some less-expensive valves are even being considered sacrificial in applications where severe abrasives or excessive temperatures exist.

Like automated valves, control valves should also be selected based on the actuator's compatibility with the system and the environment.

This should include power source, area classifications, physical size, seismic activity, and inherent atmospheric corrosion.

The best way to collect all the necessary information is to use a control valve data collection sheet similar to one developed by the Instrument Society of America (ISA) (see Fig. 4.1). From this information the necessary flow coefficient C_v of the valve to be used may be calculated.

Control valve terms

Before discussing the valve types used in control applications a review of control valve terms is important.

Valve capacity. Valve manufacturers have adopted a common basis for rating capacities of control valves. The yardstick accepted for this purpose is termed flow coefficient C_v and has been in use since the middle 1940s. The valve coefficient C_v is defined as

$$C_v = \frac{q}{(\Delta P/G)^{1/2}}$$

where q = volumetric flow rate through valve, gal/min (L/min)
ΔP = pressure drop across valve (including inlet and outlet losses), lb/in^2 (bar)
G = specific gravity of the flowing fluid

Stated another way, it is the number of gallons per minute of water at room temperature which will pass through a given flow restriction with a pressure drop of 1 lb/in^2. For example, a control valve which in the full open position passes 25 gal/min of water with a 1-lb/in^2 pressure drop has a maximum flow coefficient of 25. The flow is similarly determined at various increments of valve rotation, and the C_v at each increment may be calculated. A plot of these valves reveals the *characteristic curve* of the valve. The curve is made by plotting the percent of maximum travel against the percent of maximum flow. The characteristic curve thus obtained reflects the control characteristic of that particular valve and determines whether it or another type of valve is best suited for a particular application.

Flow characteristics. All rotary control valves modify the flow of a fluid that passes through the varying flow opening (orifice) as the valve's closure element (ball, plug, disc) rotates in response to a control signal. Depending upon the design of the valve the change in flow may be defined in one of the following ways:

Specification Forms for Process Measurement and Control Instruments, Primary Elements and Control Valves

PROJECT ____	DATA SHEET ____ of ____
UNIT ____	SPEC ____
P.O. ____	TAG ____
ITEM ____	DWG ____
CONTRACT ____	SERVICE ____
MFR SERIAL* ____	

Fluid ____ Crit Pres PC ____

SERVICE CONDITIONS

	Units	Max Flow	Norm Flow	Min Flow	Shut-Off	
Flow Rate					—	
Inlet Pressure						
Outlet Pressure						
Inlet Temperature						
Spec Wt/Spec Grav/Mol Wt					—	
Viscosity/Spec Heats Ratio					—	
Vapor Pressure P_V					—	
*Required C_V					—	
*Travel	%				0	
Allowable/*Predicted SPL	dBA	/	/	/	—	

LINE

Pipe Line Size & Schedule | In ____ | Out ____
Pipe Line Insulation ____

VALVE BODY/BONNET

*Type ____
*Size ____ ANSI Class ____
Max Press/Temp ____
*Mfr & Model ____
*Body/Bonnet Matl ____
*Liner Material/ID ____
End Connection | In ____ | Out ____
Flg Face Finish ____
End Ext/Matl ____
*Flow Direction ____
*Type of Bonnet ____
Lub & Iso Valve ____ Lube ____
*Packing Material ____
*Packing Type ____

TRIM

*Type ____
*Size ____ Rated Travel ____
*Characteristic ____
*Balanced/Unbalanced ____
*Rated C_V ____ F_L ____ X_T ____
*Plug/Ball/Disk Material ____
*Seat Material ____
*Cage/Guide Material ____
*Stem Material ____

SPECIALS/ACCESSORIES

NEC Class ____ Group ____ Div ____

ACTUATOR

53 *Type ____
54 *Mfr & Model ____
55 *Size ____ Eff Area ____
56 On/Off ____ Modulating ____
57 Spring Action Open/Close ____
58 *Max Allowable Pressure ____
59 *Min Required Pressure ____
60 Available Air Supply Pressure:
61 Max ____ Min ____
62 *Bench Range ____ / ____
63 Act Orientation ____
64 Handwheel Type ____
65 Air Failure Valve ____ Set at ____
66 ____

67 Input Signal ____

POSITIONER

68 *Type ____
69 *Mfr & Model ____
70 *On Incr Signal Output Incr/Decr ____
71 Gauges ____ By-pass ____
72 *Cam Characteristic ____
73 ____

SWITCHES

74 Type ____ Quantity ____
75 *Mfr & Model ____
76 Contacts/Rating ____
77 Actuation Points ____
78 ____

AIR SET

79 *Mfr & Model ____
80 *Set Pressure ____
81 Filter ____ Gauge ____
82 ____

TESTS

83 *Hydro Pressure ____
84 ANSI/FCI Leakage Class ____
85 ____
86 ____

Rev	Date	Revision	Orig	App

Figure 4.1 ISA control valve sheet. (*Courtesy of Instrument Society of America.*)

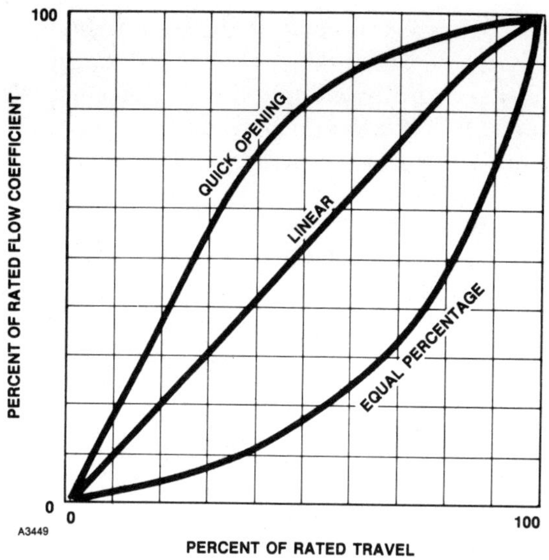

Figure 4.2 Graph of valve flow characteristics.

Equal percent. Equal percentage is an inherent flow characteristic in which a given percentage change of valve travel will produce an equal percentage change in the existing flow coefficient (see Fig. 4.2).

Linear. A linear flow characteristic is an inherent characteristic where equal increments of travel produce equal increments of flow change at a constant pressure drop.

Quick opening. A valve with a quick opening flow characteristic provides a maximum change in flow rate at low travel. These valves are often used where significant flow rate must be established quickly as the valve begins to open.

Plant experience and analysis indicate that if there is doubt about what flow characteristic to choose, a valve with equal percentage flow is the better choice. Fortunately most common ball, plug, and butterfly valves are inherently equal percentage, while high-performance butterfly valves are closer to linear.

When other than the inherent flow of a valve is required for a particular application, characterized cam positioners are often used. These positioners feature a cam and position feedback mechanism that allows variations in the valve position at a given input signal.

Rangeability. The maximum flow capacity of a control valve is provided in the full open position. This is, of course, when the flow opening is at its maximum. Optimum control—or regulation—of the flow through a control valve is exhibited in a range that is generally between 30 to 70 percent of rotation. Full-port ball valves give sluggish performance beyond 70 percent, but reduced bore designs have increased resistance to flow and allow good throttling characteristics to near full valve capacity.

Rangeability of a control valve may be defined as the ratio of maximum controllable flow to the minimum controllable flow. Stated another way, it is the ratio of maximum to minimum *usable* sizing coefficient. It is obtained by dividing the minimum usable sizing coefficient in percent into the maximum usable sizing coefficient in percent.

Rangeabilities of control valves vary, depending somewhat on the body style. Throttling ball valves may have rangeabilities in excess of 100:1; standard butterfly valves have rangeabilities between 10:1 and 20:1; and pinch and diaphragm valves have rangeabilities as low as 5:1.

High recovery. In some valve designs, particularly butterfly valves, the downstream pressure can recover to a pressure substantially higher than the pressure at the vena contracta, or smallest cross-sectional area, of the flow stream. This high-recovery characteristic may lead to valve damage through cavitation and should be considered.

Cavitation. Cavitation is defined as the noisy and potentially damaging formation and collapse of vapor cavities formed when the pressure of a liquid drops below its vapor pressure at the vena contracta and then recovers to a pressure level above its vapor pressure downstream of the control valve. Since rotary valves are generally considered high-recovery valves, one should be aware of their cavitation potential. Most modern sizing techniques will indicate when cavitation potential exists.

Pressure drop ratings. Sometimes more limiting than the ANSI class rating is the pressure drop rating. Pressure drop is defined as the difference between the inlet and outlet pressures of the valve. Pressure drop capabilities are generally a function of a valve's internal parts: closure members, shafts, bearings, and so forth. The ability of a ball, butterfly, and other rotary shaft valves to handle pressure drop is largely dependent upon the support of the throttling member, the bearing sizes and locations, and the sealing member design. Consequently, there is considerable variance in pressure drop capabilities between these common rotary-shaft valves. A determination should be made before the final selection.

Standard rotary valves for control

Depending upon the degree of precision and the severity of the control application, standard on-off rotary valves are often adequate for flow control. They are economically priced, small, and lightweight and also provide bubble-tight shutoff if required. Though each of the most popular rotary valves (ball, plug, and butterfly) have various design features, a basic understanding of their performance should be useful for process control selection.

Because flow control and general performance characteristics of a rotary valve are basically functions of how well the closure element permits or restricts flow based on the process requirements, our discussion of the various types of control valves will focus primarily on how flow is achieved by each. Where performance characteristics are important, they shall be included in this text.

Ball valves. Floating ball designs use upstream pressure to force the ball into the seat for shutoff. Therefore, the drive shaft and ball must be loosely connected so that the ball is free to seat. At the same time the shaft must be able to transmit positive motion to the ball. This is often accomplished using a drive tang. The tang is arranged so that the ball is free to move toward the seat when the valve is closed.

To prevent erratic control due to deadband, the shaft and ball (or disc) should function as a single unit. For example, if a ball is loosely connected to the shaft, there is no response to a small change in control signal. Some manufacturers offer a matched ball and stem to minimize the machined tolerance in the fit (see Fig. 4.3).

Ball valves exhibit equal percentage flow characteristics which is a natural result of the orifice configuration. Some of the advantages of the ball valve in control are:

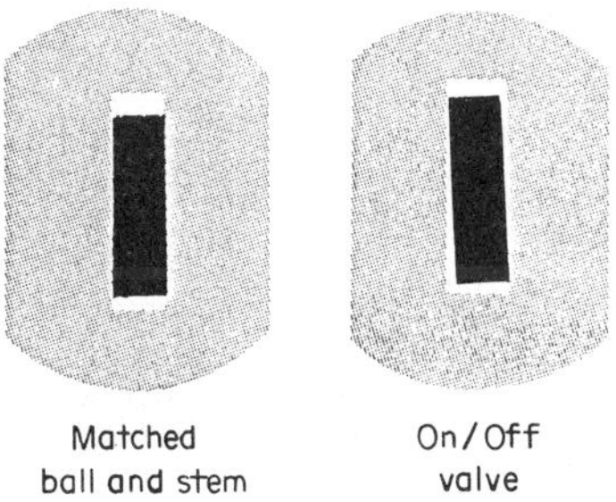

Figure 4.3 Matched slot and stem improves control performance for ball valves.

- It is not subject to hydrodynamic torque as are butterfly valves.
- Its C_v is generally higher than for equivalent globe valves.
- It is available in a wide range of sizes.
- It provides two stages of control as the fluid passes around the inlet and outlet edges of the ball port opening.

The ball control valve will handle a wide spectrum of control applications. There are, however, limiting conditions under which the valve should not be utilized.

The valve should not be used when the temperature of operation is above 500°F (260°C), for control of liquid flow at velocities greater than 15 ft/s (5 m/s), or in flow systems requiring a C_v value less than 0.25.

Butterfly valves. There are several types of butterfly valves that may be used in control service. They range from inexpensive unlined or rubber-lined valves to high-performance butterfly valves, which give high pressure drop capability and good shutoff. By far the most popular is the inexpensive rubber-lined valve. Butterfly valves are high-recovery valves and are sensitive to cavitation. However, strict adherence to the manufacturer's sizing data usually eliminates the cavitation problem. Further, butterfly valve torque is sensitive to pressure drops. Fluid flow across a butterfly valve disc creates lift or drag similar to that created across an airplane wing. Since the position of the valve disc changes when the valve is opened or closed, the amount of lift or drag varies greatly. The valve actuator has to move and hold the disc in its proper position. If the actuator output does not meet the valve torque requirements, erratic control results. Offset-disc butterfly valves are less susceptible to these effects, but for severe control, contouring of the disc may be required.

Butterfly valves are not typically available in sizes under 2 in (50 mm) because the disc profile takes up such a large percentage of the flow area. As a result, the capacity of these small sizes is limited. In the fully open position, butterfly valves offer little resistance to fluid flow and dissipate relatively little flow-stream energy due to streamlined internal contours and minimal flow interruption. When compared with other control valve types, conventional butterfly valves have higher capacities per dollar. This cost advantage increases dramatically in sizes over 12 in. They are often preferred for very large control valves.

Valves with conventional discs have a rangeability of 15:1 and are limited to controlling the flow 15 to 60° open. Discs have been designed to extend the range by controlling the flow to 90° or fully open.

The valve C_v should not be less than the sizing C_v when the valve is 60° open.

Plug valves. Being a rotary valve, the sleeved plug valve controls flow similar to a butterfly or ball valve. As the valve plug is rotated from 0 to 90°, the size of the flow passageway is increased from 0 to 100 percent. At low degrees of opening, the flow through the valve is somewhat tortuous and is therefore similar to a globe valve. However, after the valve is opened approximately 50 percent, the flow becomes less turbulent through the valve. This more efficient flow passage means the valve exhibits high-recovery characteristics. Consequently, a large C_v per square inch of flow area is obtained. The combination of the increased efficiency of the valve at larger degrees of opening and the throttling effect at low degrees of opening due to the tapered plug means that the valve has an inherent equal-percentage-type characteristic. These same two factors also explain why the rangeability of the TFE sleeved plug valve is in excess of 100:1, depending on size. Other characteristics can be obtained by changing the basic shape of the flow opening in the plug or in the cage.

Because of the interference fit between the plug and valve sleeve, there is a substantial friction requiring large torques to position the valve. However, this friction makes the valve extremely stable and vibration resistant as fluid forces are exerted on the plug. The effect of unbalanced forces at any throttling position is negligible.

Characterized rotary valves for control

When standard rotary valves are not adequate for control service because of flow characteristics, pressure drop, or even material compatibility, some manufacturers offer specially engineered versions of their valves to satisfy special requirements. These characterized valves include segmented balls, V-port plugs, specially contoured discs for butterfly valves, and even erosion-resistant material such as ceramic. Often more costly than their standard counterparts, these valves play an important role in total valve selection for the process engineer.

V-notch ball valves. The V-notch ball segment valve uses only a portion of a spherically shaped component (partial ball) to control flow (see Fig. 4.4). The segment remains in contact with the sealing surfaces throughout its rotation. The unrestricted, straight-through flow design provides high capacity, while the V-notch provides exceptionally wide rangeability (approximately 300:1). In addition, a shear-on-

Figure 4.4 Cutaway of V-notch ball valve. (*Courtesy of DeZurik.*)

closing action makes this valve suitable for use with fibrous slurries and fluids with entrained solids.

The geometry of the V-notch ball segment combines with the straight-through flow path to provide wide rangeability to accurately control both very low flow rates and very high flow rates. Flow is controlled from the point where the notch of the ball just begins to expose the port to the fully open position. The open flow path of this valve design also prevents mixtures or slurries from settling out. V-notch balls are commonly designed to provide an equal percentage flow characteristic which is optimum for a broad range of applications. Other flow characteristics can be developed by changing the shape of the notch.

V-notch valves are often used in the pulp and paper industry as well as in sewage treatment plants because of their shearing action and ability to control heavier, thicker media. Their wide rangeability makes them appropriate for use in applications requiring throttling precision over a broad range of flow rates, such as in basis weight control or in certain blending or mixing processes. Models with stainless-steel or TFE stainless-steel seals are often selected for use with steam and hot water over 200°F (93°C), anhydrous ammonia, alkalis, caustics, pulp and paper liquor services, and other demanding services.

V-seat ball valve. V-seat control valves incorporate a characterized metal seat with a standard quarter-turn ball valve design (see Fig. 4.5). The V-seat control valve is adaptable to a range of pressure and control applications generally unachievable with soft-seated rotary control valves. These valves are used in industries such as chemical processing, pulp and paper, and mining and can be installed anywhere V-ported ball valve types are used.

The characterized 30° V-ported metal seat may be made of graphite-impregnated stainless-steel alloy and offers abrasion resistance, temperature capability to 1000°F (537°C), and pressure capability to 1000 lb/in^2 gage (69 bar). The V-ported metal seat is capable of performing in high-pressure-drop applications where standard rotary ball valves do not last and exhibit seat foldout. With the metal seat downstream and resilient or optional metal seat upstream, these valves exhibit

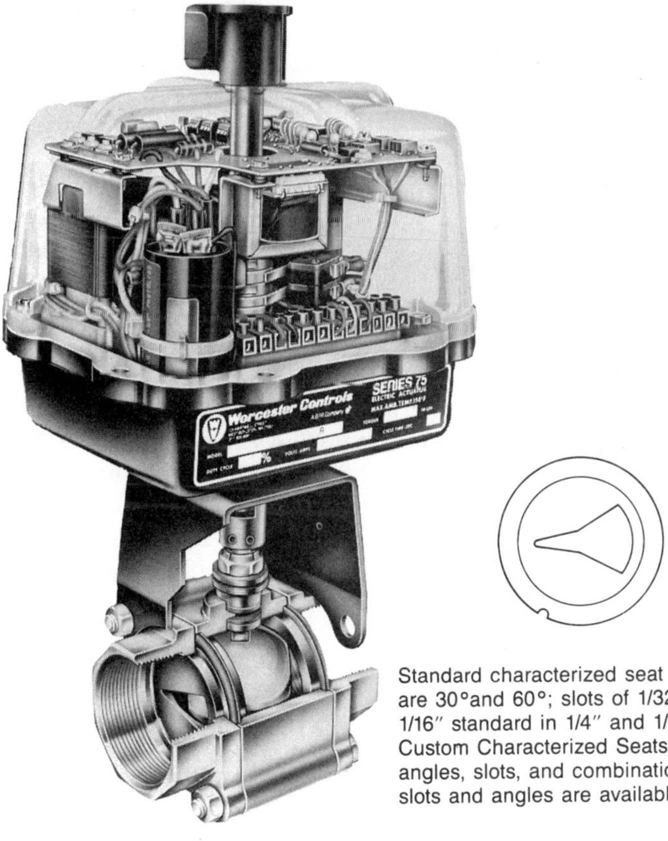

Standard characterized seat angles are 30°and 60°; slots of 1/32″ and 1/16″ standard in 1/4″ and 1/2″ valves. Custom Characterized Seats with other angles, slots, and combinations of slots and angles are available.

Figure 4.5 Characterized V-seats are available with various angles of opening. (*Courtesy of Worcester Controls.*)

high-pressure-drop capability, accurate control, and meet the shutoff requirements of ANSI B16.104, class VI.

Eccentric rotary plug valves. The eccentric rotary valve incorporates design and construction features which make it particularly suitable for highly erosive applications (see Fig. 4.6). Streamlined flow passages, rugged components, and many choices for erosion-resistant trim materials are features which contribute to long life.

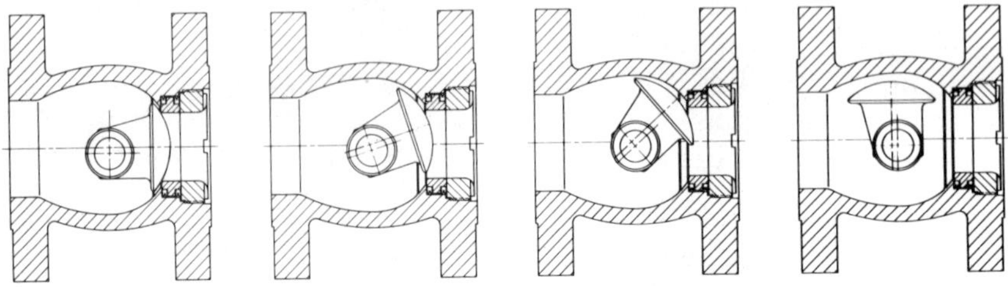

Figure 4.6 Closing action of eccentric plug control valve. (*Courtesy of Fisher Controls International, Inc.*)

Contours of the valve body are designed to direct erosive flow away from critical internal surfaces to minimize erosion damage and reduce particulate buildup. This valve style is generally available in a variety of corrosion- and erosion-resistant materials to further extend valve life. Materials with increased erosion resistance may be incorporated in components where erosive activity is focused.

The valve seat design uses a solid metal shutoff surface without thin elastomer or metal seals to erode in service. The seat is held in position by a retainer but allowed to float. The seat ring is centered by the action of the plug closing against it. This self-centering feature eliminates many alignment problems in manufacture and maintenance. The floating seat also provides a self-lapping action between the plug and seat, which promotes longer effective plug and seat life.

Cage control plug valves

Cage control plug valves provide soft seat protection in throttling applications and in services with high pressure drop through the valve (see Fig. 4.7). Caged plugs allow metal-to-metal throttling and metal-to-soft-seat shutoff for both automatic and manually controlled applications. The cage is located within the plug and is fixed in the valve body. PTFE bearings on the flow cage allow the plug to rotate freely.

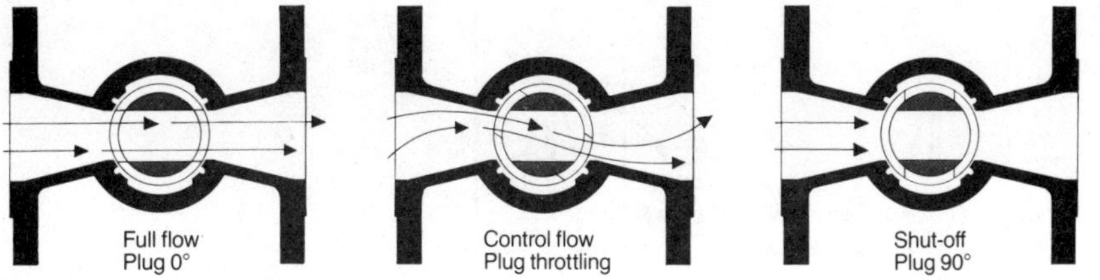

Figure 4.7 Closing action of cage control plug valve. (*Courtesy of Xomox Corporation.*)

The cage protects the seat from erosion in throttling applications by shielding the soft seat from direct flow impingement at the valve ports. The cage mechanism also serves to reduce turbulence and the cutting action of high-velocity liquids, slurries, and gaseous vapors by providing a more direct, contoured flow path through the valve.

Caged plugs are available in 1 through 8 in (25 to 200 mm) sizes in the following standard materials: WCB body, CD4MCu plug and cage; 316SS body, CD4MCu plug and cage; A20 body, A20 plug and cage. Other sizes and materials are also available.

Dynamically balanced disc butterfly valves

Pressure drop across a butterfly valve disc varies for a given flow rate and valve disc position. This variation generates a resulting hydrodynamic torque on the valve shaft. The magnitude of this torque varies as the valve disc is rotated. If actuator torque output is not sufficient to meet the changing torque requirements of the valve, erratic control will result.

Conventional discs may be rotated from 0 to 90° for on-off service but are generally limited to 60 to 70° maximum rotation for throttling service. The throttling limitations of standard discs are the result of dynamic fluid forces creating excessive torque instability between approximately 60° and the fully open position.

Dynamic forces created by fluid flowing around conventional disc designs increasingly tend to close the valve as it rotates between roughly 60 and 70° open, then rapidly diminish as the valve travels beyond approximately 75°. These changing torque requirements can make actuator sizing and controller tuning difficult.

The magnitude of the dynamic torque peaks may be lowered by replacing the conventional disc with a dynamically balanced disc (see Fig. 4.8). This disc design is dynamically balanced to even out the varying pressures along portions of the disc's surface as it rotates. The shape of the disc protrudes into the flow stream and decreases flow and velocity on the disc's downstream edge. This causes increased flow and pressure on the leading edge, so the tendency for process flow to

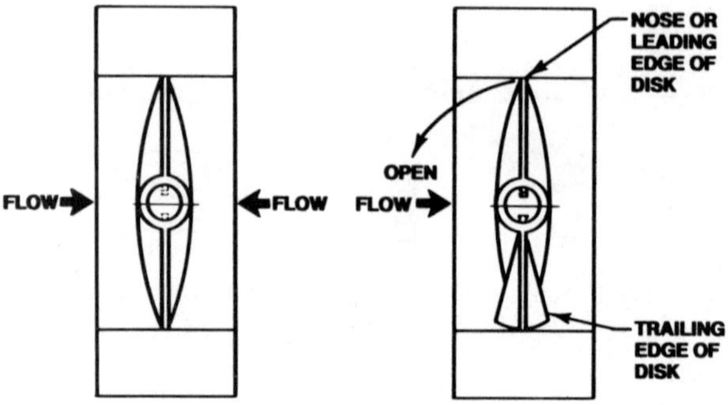

Figure 4.8 Standard and contoured disc butterfly valve. (*Courtesy of Fisher Controls International, Inc.*)

close the valve is greatly diminished. By balancing the flow separation so that equal amounts of flow go around each side of the disc, velocity and pressure are roughly balanced. The geometry of the disc is usually calculated so that a small tendency to close the valve remains.

Selecting a control valve

The following information is necessary to consider when choosing a rotary control valve.

Service conditions

1. Fluid
2. Flow rate minute and flow rate maximum
3. Flow rate normal
4. Pressure in (P1) and pressure out (P2)
5. Sizing pressure
6. Temperature maximum and normal
7. Specific gravity (at flow temperature)
8. Viscosity
9. Line size

Valve

10. Size
11. Pressure rating
12. Maximum temperature

Pneumatic actuator and positioner

13. Air pressure available
14. Action
15. Failure position
16. Actuator size
17. Positioner type
18. Control signal [(3 to 15 lb/in^2, etc.) (.2 to 1 bar)]
19. Standard and reverse acting
20. Characteristic

Electric actuator and positioner

21. Enclosure rating (weatherproof, hazardous, etc.)
22. Voltage supply
23. Cycle time (0 to 90°)
24. Control signal (4 to 20 mA dc, etc.)

Accessories and remarks

____

____

____

In general, the selection of a control valve requires three steps: controllability, compatibility, and actuation. Controllability means the ability of the valve to handle the pressure, temperature, flow rate, and medium. This leads to the flow characteristics, noise, cavitation, and flow capacity.

Compatibility is basically material selection. Corrosion, excessive temperatures, and even the requirement for bubble-tight shutoff all require careful material selection. Some users tend to overspecify shutoff requirements. Few control valves are actually required to perform the function of a block valve.

Actuation is important for performance as well as system compatibility. Pneumatic and electric actuators are capable of providing long, accurate service life if properly selected and applied. Positioners are available for unfiltered air, harsh environments, hazardous areas, and even with specially engineered cams for unique needs.

Of course the cost of a control valve both initially and over its performance life should always be considered. Overspecifying is a conservative means of ensuring control, but this may lead to a high purchase price, wasted energy, and excessive operating costs.

Actuator selection and sizing should involve the experience of both the user and the manufacturer. Later in this book you will find extensive information on actuator performance and selection and the accessories used to communicate with the control system.

4.2 Valve Design Options

Multiported valves and tee assemblies

An especially unique feature of rotary-shaft valves is that they may be manufactured as multiported devices. A multiported valve is used to divert or distribute a fluid, for isolating equipment while staying "on-stream" or for blending two different fluids. An example of the latter would be a tee assembly where two butterfly valves are mounted to a common tee and are controlled by a single actuator. Hot and cold water flow into the tee and are blended to provide a desired output water temperature. This tee arrangement is used extensively in heating, ventilating, and air conditioning (HVAC) systems for controlling building temperature.

When selecting a multiported valve the following should be considered:

1. Valve type
2. End connections
3. Flow arrangement
4. Need for bidirectional shutoff
5. Number of positions

Most ball and plug valve manufacturers offer multiported valves. By far the most popular number of ports is three. This type of multiport valve is used primarily for diverting fluids. A greater awareness of the available porting arrangements, however, may lead to more creative, economical, and efficient piping system design.

Three-way valves. At the heart of any multiported valve is the design of the diverting element. This is what provides the flow pattern and shutoff capability and the ultimate versatility in the number of flow pattern arrangements.

When a three-way valve is described, its basic two-way arrangement serves as the reference. For instance, if a two-way valve has a third port added to its bottom (the location opposite of the valve shaft),

Figure 4.9 Side-ported three-way ball valve. (*Courtesy of KTM, Inc.*)

it is referred to as a bottom-ported three-way valve. If the port is added to the side of the two-way body it is referred to as a side-ported three-way valve (see Fig. 4.9). To accomplish three-way flow patterns the ball or plug is manufactured with an L, double L, cross, or T port.

When used for simple diverting applications, a three-way valve usually requires only 90° of rotation. A common practice in describing the flow arrangement of a three-way valve is to label the ports *A, B,* and *C*. *A* and *B* are used to identify the basic two-way valve ports and *C* identifies the additional third port.

When only 90° of rotation is required, the valve is referred to as a two-position, three-way valve. This also applies to valves that are used to provide a flow pattern when 0 to 180° of rotation is required. However, most 0 to 180° three-way valves pass through a 90° porting arrangement that is sometimes in a flow scheme. In those cases the valve is referred to as a three-position, three-way valve. The number of ports and flow possibilities is almost unlimited. Some manufacturers offer four-, five-, and even six-way valves (see Fig. 4.10). Using these available patterns valves may be interconnected to create as much as 24-way flow arrangements.

Tee assembly. As mentioned earlier, standard cast iron (or stainless steel) tees may be fitted with two valves (most often butterfly valves)

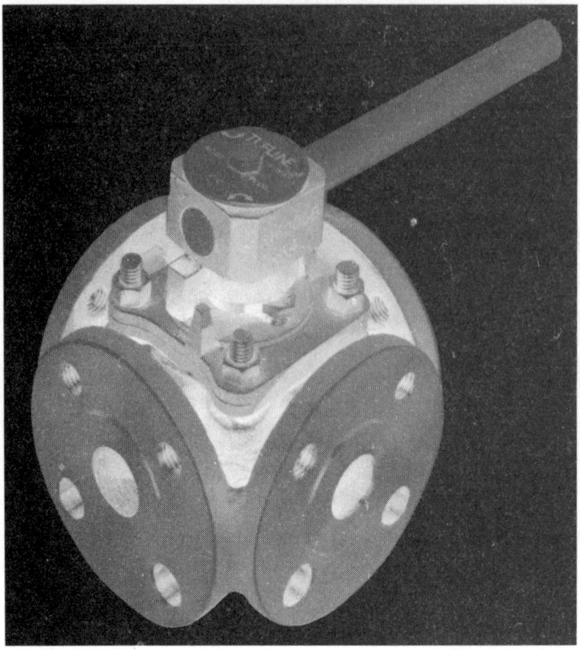

Figure 4.10 Four-way plug valve. (*Courtesy of Xomox Corporation.*)

to create a desirable flow diverting or blending assembly. These valves are usually operated by a single actuator for on-off or modulating control. When specifying a tee assembly the valve location on the tee should be indicated (see Fig. 4.11). Tee assemblies are practical in line sizes up to 12 in (300 mm).

Sanitary valve design. For our purposes sanitary valves will be defined as valves safe for processes where an environment free of viable bacteria, etc., must be maintained. This may involve liquids, semiliquids, and solids. Sanitary valves employ a design that maintains product security while allowing reliable sterilization of the critical valve areas prior to processing. Sanitary valves are available in ball, plug, and butterfly designs. The design concept of one sanitary valve manufacturer includes a two-chamber sealing system that safeguards against biocontamination from both external and internal sources (see Fig. 4.12). A shaft-sealing system separates the disc and liner areas from the atmosphere and from the valve's internal steam chamber. This system remains bubble-tight even under vacuum conditions.

A second internal sealing system creates a self-contained steam sterilization chamber. This chamber permits the passage of steam

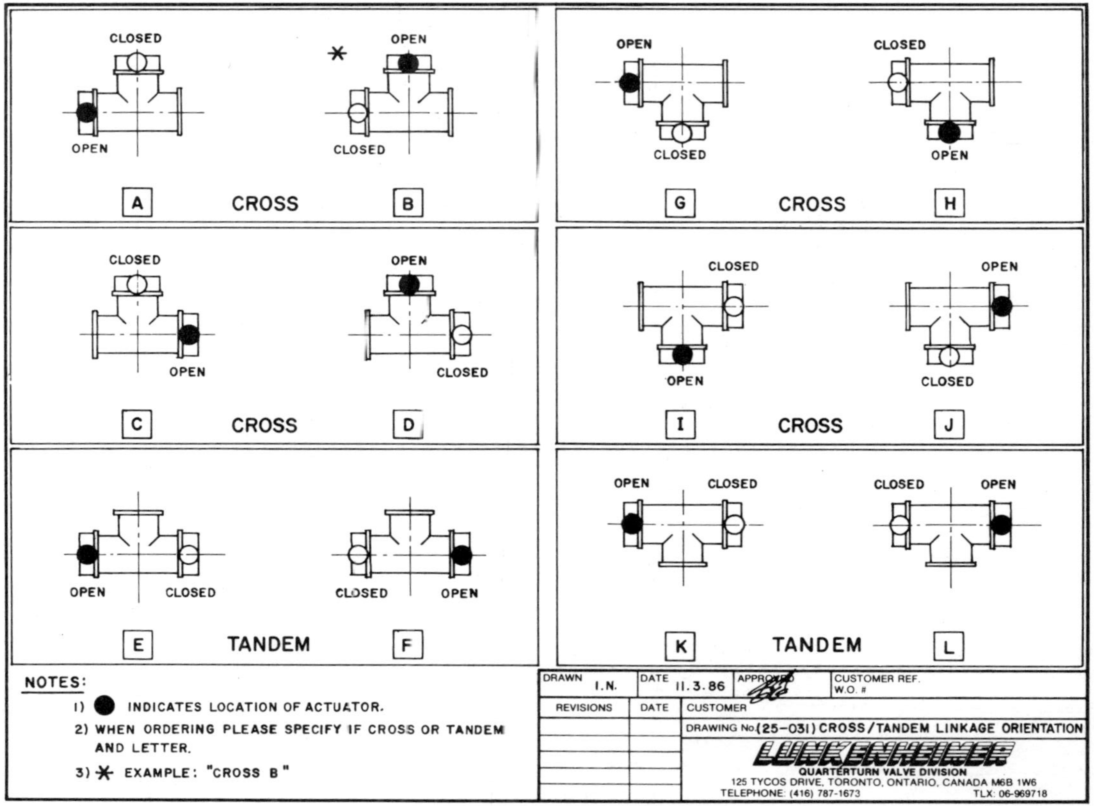

Figure 4.11 Configurations and valve locations for tee assembly. (*Courtesy of Lunkenheimer Canada.*)

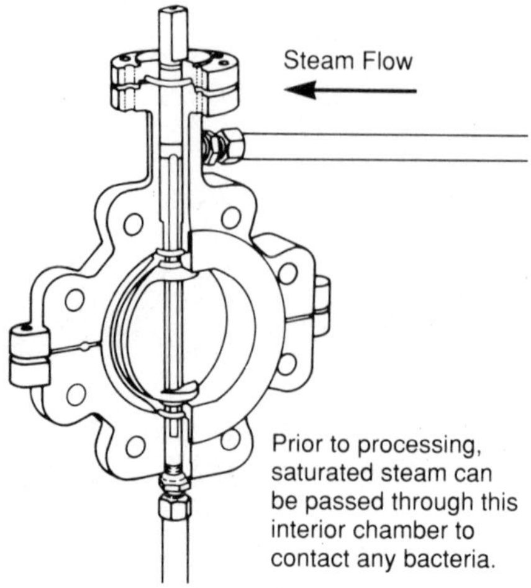

Figure 4.12 Sterilizing system for sanitary butterfly valve. (*Courtesy of Garlock, Inc.*)

through the critical internal areas that could not be reached by traditional steam flushing methods. Prior to processing, saturated steam can be passed through the liner and disc areas as well as through the internal steam chamber. This two-chamber steam flushing capability assures sterilization.

Sanitary valves are available in the following materials and pretreatments: cast stainless steel, type 316L; cast stainless steel, type 304L; ultrasonically cleaned. Other materials are available.

A Food and Drug Administration (FDA)-approved PTFE or rubber liner is often specified in sanitary service. Some are isotactically molded to achieve high density and crystallinity. The liner should be void-free, with a minimum of 0.125 in (3 mm) thickness at all points. Stainless-steel discs for such valves are available in cast or forged 316L stainless steel, mechanically polished and/or electropolished to provide a very smooth surface that is virtually free of all surface irregularities. This discourages bacterial growth.

Cryogenic valves. Cryogenic valves are used to control fluids in a very low temperature range, between −280 and −445°F (−173 and −265°C) (100 and 8 K). The majority of the valves used in cryogenic work handle gases or fluids at comparatively low pressures, less than 500 lb/in^2 gage (35 bar). Selection with regard to pressure is either class 150 or 300 (PN 20 or 50).

Selection with regard to materials is more exacting. The valves may be divided roughly into three categories:

- Warm valves, which are part of the operating process but are located upstream of the cold box

- Cryogenic isolation (block) valves, which are fully open or fully closed
- Cryogenic process control valves, which may be either pneumatically or electrically operated

The selection of process control valves depends on the

- Type of control required, such as flow control, pressure reduction or throttling, and isolation
- Type and degree of automation in order to obtain the best combination of valve and actuator
- Operating conditions, such as liquid or gaseous phase, and temperatures
- Individual preference for facility of inspection and maintenance versus initial valve costs

All cryogenic process valves operating within the insulated casing (cold box) should have an extended bonnet and stem protruding from the casing so that the handwheel and/or the actuator are located outside, and under ambient conditions (see Fig. 4.13). Automated cryogenic valves should have a manual override for emergency purposes.

In general, the materials of construction for body, bonnet, stem, and disc or plug should be either 316 stainless steel, K-Monel, manganese bronze, or even aluminum (excluding stem and plug). The packing should preferably be made of PTFE and be of the V-ring type, and the gasket should also be made of PTFE.

Within the insulated casing where cryogenic process equipment is housed, it is important to have zero external leakage. For this reason, butt-welded valve connections are often used. Transition couplings are readily available on the market to permit the welding of stainless-steel valves to piping of another material such as aluminum.

Stainless-steel valves are also available with top-entry bodies so that the body itself does not have to be removed for servicing. The bonnet can be unbolted allowing the stem, disc or plug, and trim to be removed for inspection and maintenance.

4.3 Unique Valve Designs

To satisfy the valve performance requirements of a variety of industries most manufacturers will offer unique versions of their valves if requested. Some of these designs have led to standard products now being offered for wider use. A sampling of some unique valve designs is discussed here.

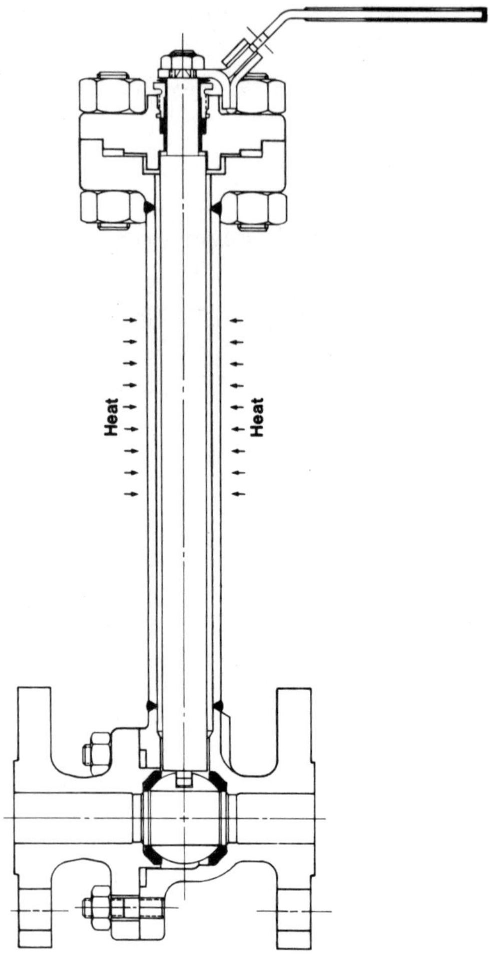

Figure 4.13 Extended bonnet for cryogenic service valve. (*Courtesy of Velan Inc.*)

Rising stem ball valves

The rising stem ball valve combines the operating principle of the threaded stem gate valve with the rotary-cam action of a ball (see Fig. 4.14). This valve uses the mechanical loading advantage of the threaded stem to create bubble-tight sealing around a ball. As you begin turning the handwheel to open the valve, the trunnion-mounted core (ball) is mechanically pulled away from the seat. This is accomplished by the camming action of the lower part of the valve stem as it moves upward. Only after the core is clear from the seat does it begin turning to the open position. The rotation of the core is caused by a spiral groove allowing the stem and therefore the core to turn 90° at

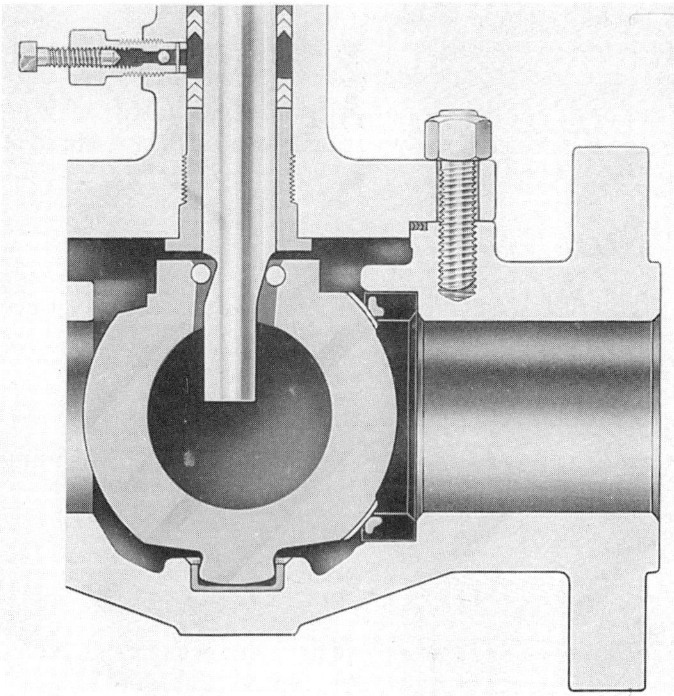

Figure 4.14 Rising stem cam action ball valve. (*Courtesy of Orbit Valve Company.*)

the proper point in the cycle. This action eliminates the friction that eventually causes wear in other valves. When open, the flow is straight through reducing turbulence.

To close the valve, the handwheel is turned in the opposite direction and the core starts to turn. The final turns of the handwheel lower the stem, tilting the core forward, where it is mechanically cammed against the seat. The result is a mechanically energized, positive seal. These valves are used in low- and high-temperature, fire-safe, and block-and-bleed applications and are available with metal or soft seats.

These valves are ideal for automation by the very nature of their operating principle. Many automated valves remain open or closed for extended periods of time. Environmental and process conditions create situations requiring higher breakaway torque in valves which creates friction by scraping the valve seating surfaces together. A rising stem ball valve pulls the core away from the seat with a mechanical camming action to break the mechanically energized seal. The core is then rotated to the open position without scraping the seat. This reduces the need for high-torque actuators found on many other multi-turn valves which scrape their seating surfaces together during opening and closing.

By eliminating opening and closing friction these valves may be supplied with low-pressure pneumatic actuators and still be assured of sufficient thrust to open and close.

Rising stem ball valves (summary)

Sizes: 1 to 20 in (25 to 500 mm)

End connections: Flanged, threaded, welded

Body materials: Steel and stainless steel

Seat materials: Nylon, TFE, stainless steel

Temperature range: −150 to 800°F (−100 to 425°C) (depending upon materials)

Pressure range: up to 2500# ANSI (PN 400)

Inflatable-seat butterfly valve

Rubber-seated elastomer-lined butterfly valves are economical valves for the control of liquids, gases, and solids. Shutoff is achieved when the valve disc and seat are aligned with an interference fit. In liquid flow control applications the disc-to-seat contact is lubricated by the fluid and helps minimize the operating torque as well as wear on the seat. In gas and solids applications where the media adds no lubrication to the seat area the operating torque is higher because of the friction and rubbing between the two surfaces which results in premature seat wear. For valve applications handling gases and dry bulk solids a butterfly valve with an inflatable seat may be considered (see Fig. 4.15).

How it works. The inflatable-seat butterfly valve is designed with a slight clearance between the disc and liner (seat). Because there is no contact there is no wear on the seat as the disc is rotated to the closed position and there is very little operating torque which results in smaller actuators being used than those for traditional butterfly valves. Once the disc is in the properly closed position air pressure is applied behind the specially designed liner. The liner inflates and makes contact with the disc to effect shutoff. The sequence of these events is accomplished with a pneumatic actuator and control circuit to ensure that the disc is properly aligned before the seat is inflated. The additional cost of the controls is offset somewhat because smaller actuators may be used because of the lower operating torque. The

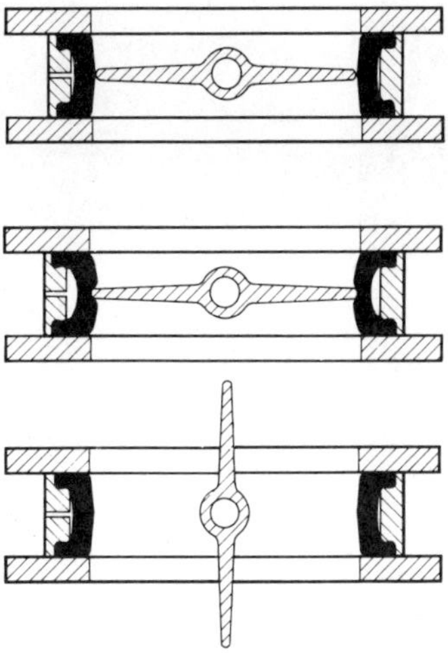

Closed, unsealed. As the valve rotates into the closed position, the disc makes only casual contact with the seat, reducing friction, wear and torque requirements.

Closed, sealed. After the valve is closed, the seat inflates against the disc providing more sealing surface and an even pressure distribution against the disc.

Open, unsealed. Before the valve opens, the seat is first deflated. The disc is then free to rotate to the open position.

Figure 4.15 Operation of inflatable-seat butterfly valve. (*Courtesy of Dynamic Air/posi-flate.*)

inflatable-seat butterfly valve may be an ideal choice in applications where frequent repair or replacement of valves in dry service is undesirable.

Inflatable-seat butterfly valves are available with the following specifications:

Sizes: 4 to 16 in (100 to 400 mm)

Body style: Wafer

Seat material: Polyurethane, BUNA-N, and FDA

Disc material: Ductile iron, 316 stainless steel

Body material: Ductile iron

Temperature range: varies with materials

Pressure range: up to 100 lb/in^2 gage (7 bar)

Nonspherical ball valve

Where floating ball valves depend upon line pressure for sealing, this valve uses a nonspherical ball plus a beveled edge on the lip of the port opening (see Fig. 4.16). The bevel minimizes pressure on the seats when the valve is open and reduces operating torque. When the valve

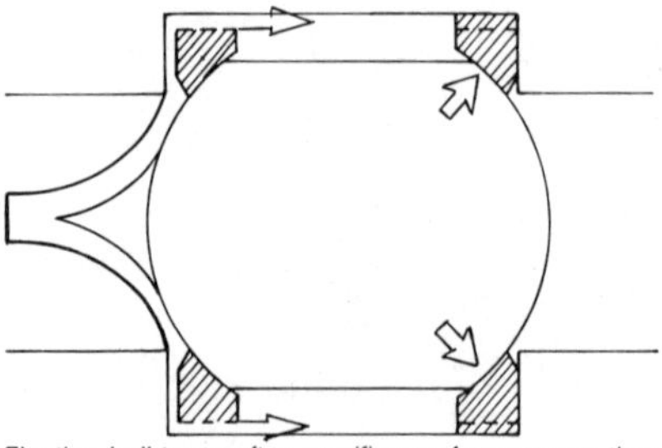

Floating ball types often sacrifice performance as they depend upon line pressure for a dependable seal.

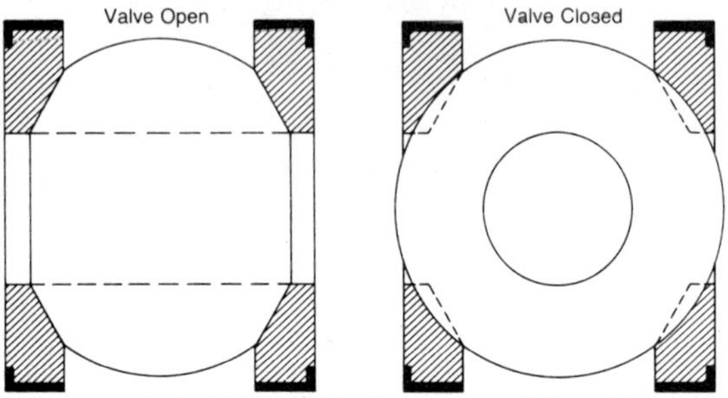

Unique non-spherical ball mechanically compresses both upstream and downstream seats to effect a tight, dependable seal independent of line pressure.

Figure 4.16 Nonspherical ball compresses upstream and downstream seats for shutoff. (*Courtesy of ITT Engineered Valves.*)

is closed, the camlike action of the ball against the seats assures shutoff independent of line pressure.

Because this valve is a low-torque valve, lateral loading on the stem seals is minimized. It also requires a smaller actuator than is necessary on conventional ball valves. Where other valves design bearing functions into their seals, this design does not require it. The stem seals are live-loaded using belleville washers.

Nonspherical valves are available with the following specifications:

Sizes: ½ to 6 in (12 to 150 mm)

End connections: ANSI 150 to 600 flanged (PN 20 to 100), ANSI 150 to 600 butt weld (PN 20 to 100), ANSI 150 to 600 socket weld (PN 20 to 100), ANSI 150 to 600 screwed (PN 20 to 100)

Body materials: 316 stainless steel, Monel, Alloy 20

Seal materials: TFE, EPDM, PEEK (polyetheretherketone), UHMWPE, Kel-F, fire safe

Temperature range: up to 550°F (290°C) (depending upon materials used)

Pressure range: up to 1400 lb/in^2 gage (100 bar) (depending upon materials used)

These valves are also available fully lined up to 6 in (150 mm) with the following materials: Tefzel or polypropylene (150# flanged only).

Diaphragm-spool valve

A combination diaphragm–pinch–quarter-turn valve is available in small sizes for hard-to-handle fluids (see Fig. 4.17). This valve provides on-off and throttling control for liquids, solids, and slurries. Like a pinch valve it utilizes a diaphragm spool (tube) that may be externally pressurized with air or hydraulic fluid to effect closure. As a manual quarter-turn valve it uses a unique double-cam mechanism which squeezes the elastomer. The valve may also be fitted with a rotary electric or pneumatic actuator with or without positioners.

Sizes: ¼ to 3 in (6 to 80 mm)

End connections: Wafer to fit between flanges

Body materials: Aluminum and special coatings

Spool materials: Standard and food grade elastomers and TFE

Temperature range: −40 to 300°F (−40 to 150°C)

Pressure range: Vacuum to 125 lb/in^2 gage (9 bar)

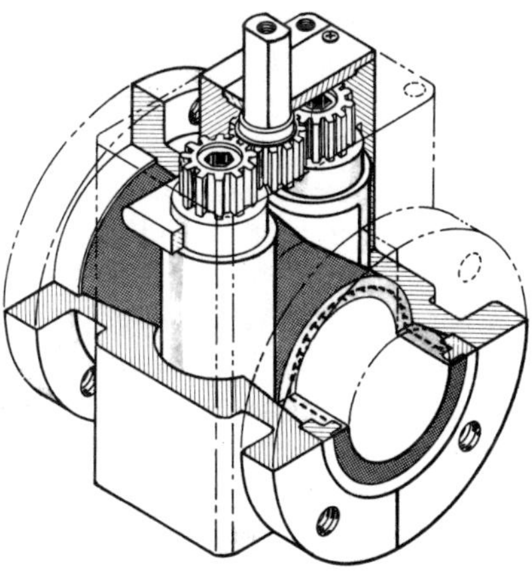

Figure 4.17 Shutoff mechanism for spool valve. (*Courtesy of Larad Equipment.*)

Solenoid-actuated control valves

Solenoid-actuated control valves are globe-bodied designs with quick change trim (see Fig. 4.18). The primary usage is in severe service applications, to the limits of a 2500# special class design.

The bonnet assembly consists of the bonnet flange, bonnet tube, and indicator tube, allowing system pressure into the uppermost regions of the valve. Central to the bonnet tube is a fixed core and movable core (plunger) making up the primary working parts of the magnetic circuit. The magnetic field of the coil permeates the pressure boundary of the bonnet tube, causing the plunger to be attracted to the fixed core.

For an on-off valve, when rated current is applied to the solenoid coil and the magnetic circuit is energized, the plunger will be attracted to the fixed core causing the pilot valve to open or close its vent port in the main disc. Once the vent port is opened or closed, the control pressure will build and the main disc will close on a fail open valve or the control pressure will vent and the main disc will open with a fail closed valve, respectively. Upon removing the solenoid current, the return spring will move the pilot and, in turn, the main disc to its normal position. Attached directly by linkage to the movable core is a magnet, which passes through the magnetic reed switches to give positive open and close indication. Standard supply voltages for on-off valves are 120 volts (V) alternating current (ac) or 125 V direct current (dc).

For modulating control applications there is a solenoid-operated valve assembly and a completely electronic valve positioner. The solenoid assembly is energized, developing a magnetic force which lifts the plunger against the force of the return spring, pulling the pilot valve off its seat in the main disc. This causes a pressure difference between the top and bottom surfaces of the main disc, allowing the main disc to follow the pilot disc by servo action.

By varying the level of current going to the solenoid assembly, the pilot valve (and, in turn, main disc) can be modulated by increasing or decreasing magnetic force. The motion of the plunger moves the core rod causing the linear variable differential transformer (LVDT) to create a voltage. This dc signal, which is an indication of actual disc position, is supplied to a valve position indicator circuit and to a comparator. The 4- to 20-milliampere (mA) input command current is converted to a dc voltage. The comparator compares the dc feedback from the LVDT to the input command voltage, which is the desired disc position, and an error signal is generated. The error is amplified, and the current applied to the solenoid is varied to correct the valve position. The totally electronic valve and positioner package enables lift characterization and control of valve stroking rates.

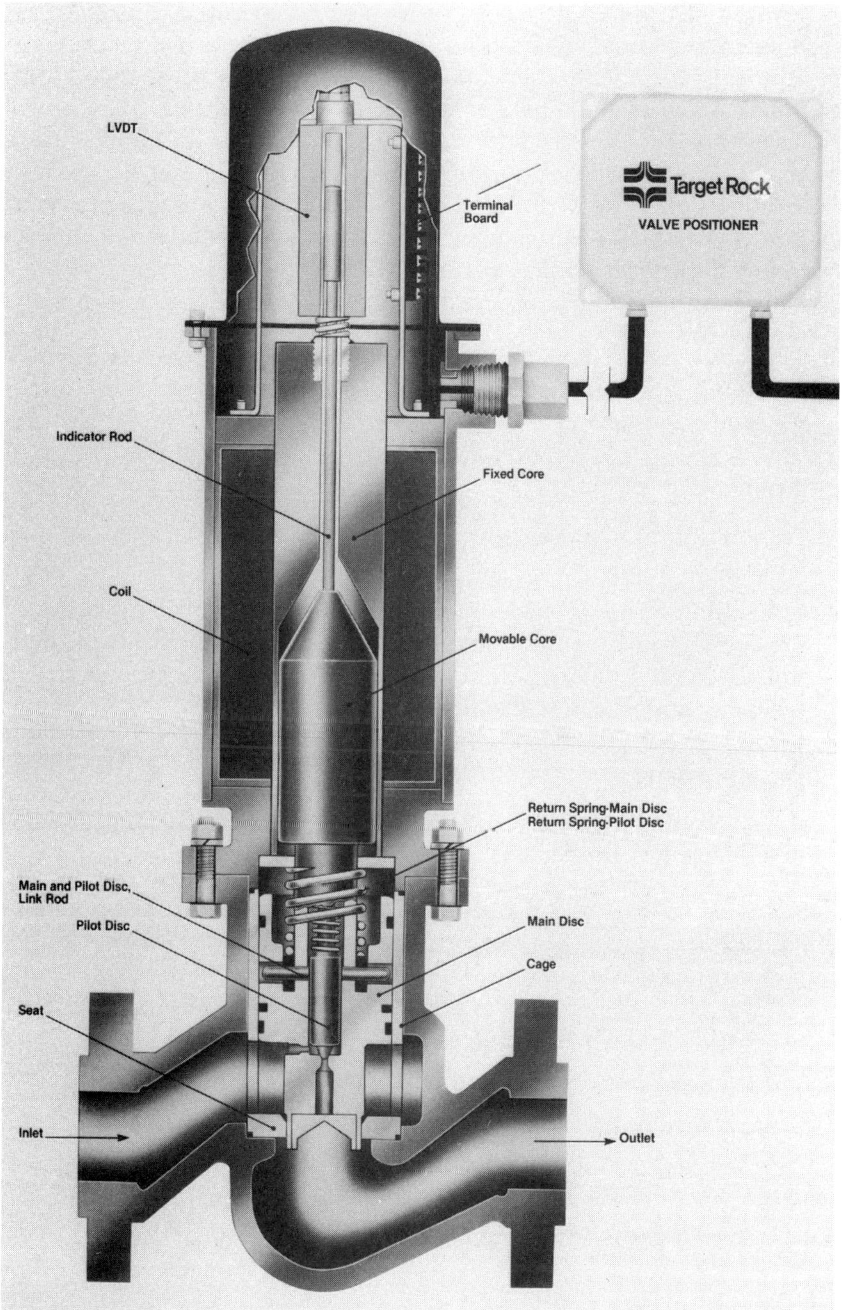

Figure 4.18 Solenoid-actuated globe control valve. (*Courtesy of Target Rock Corporation.*)

Because of unique properties of a solenoid-operated control valve, zero leakage is assured. With no rigid stem-to-disc attachment, near-perfect seat alignment may be attained. Finally, since the control chamber above the main disc has a minimal volume, the control chamber pressure vents quite rapidly. This allows the main disc to come off of its seat instantaneously. In a typical pneumatic system, it may take as much as 3 s to bleed or feed diaphragm pressure to a point where the disc will open. During this period, seat leakage, steam erosion, and deterioration may occur at the seat joint.

Some advantages to this design are

1. The ability to remote mount the positioner away from the valve due to an electronic feedback loop, reducing vibration and temperature at the positioner.
2. A 4- to 20-mA output circuit is supplied on all modulating valves, eliminating the need for auxiliary position transmitters.
3. The elimination of packing maintenance problems with no possible leakage to ambient.
4. The elimination of instrument air lines, current-to-pressure or pressure-to-current converters, air seats, air volume boosters, auxiliary solenoid valves, accumulators, mechanical positioners, and mechanical position transmitters.
5. Electronically controllable lift characteristics, opening speeds, and closing speeds.
6. An electronic, low lift, throttling limiter is used to avoid seat damage due to low lift control of cavitating fluids.

Solenoid-actuated control valves are available with the following specifications:

Sizes: 2 to 8 in (50 to 200 mm)

Materials: Steel and stainless steel

End connections: Socket and butt weld

Temperature range: up to 500°F (260°C)

Pressure range: up to 2500# ANSI (PN 400)

4.4 Valve Designs for Flammable Fluids

In many processes throughout the chemical processing and refining industries, valves and automated valves are required to handle flammable fluids. Valve manufacturers currently offer designs intended for this service called *fire-safe* valves. Selecting a valve that will provide shutoff in the event of a fire begins with an understanding of the term *fire-safe* and the standards established by users and independent testing companies.

First to consider is that a valve intended for fail-safe service will probably not be required to perform in the event of a fire for many years—and hopefully never. The valve selected, therefore, should operate as a tight shutoff valve in normal operation before a fire and during and after a fire. A review of the construction features of the various types of fire-safe valves will help in their selection.

According to the MSS, the terms *fire safe* or *fire tested* are not definitive and should not be used without an accompanying specification of what is required. Such specification may be provided in the form of a requirement for a defined test or for limitations on valve failure mode. Examples of such limitations are

1. Destruction of elastomeric materials in the valve shall not result in gross valve pressure-boundary leakage.
2. Destruction of elastomeric materials in the valve shall not result in leakage greater than the specified rate when the valve is closed.
3. External heating of the valve shall not cause uncontrolled buildup of pressure in the body cavity of a double-seated valve.

Requirements related to after-fire operability and seat tightness are difficult to define other than by actual testing using standardized procedures.

Three important criteria for evaluating fire safety in valves, and the major concerns of testing authorities, are external leakage, internal leakage, and operability.

1. Minimal external leakage. The best valve body design minimizes external leakage by eliminating large gasketed body joints and provides an adequate stem sealing arrangement of fire-resistant materials.

2. Minimal internal leakage. For fire-safe sealing integrity, some valve designs provide metal-to-metal seating prior to, during, and after exposure to a fire without relying on complete destruction of the primary resilient sealing member, or any supplementary spring loading or overtravel of the disc or ball to achieve metal-to-metal contact.

3. Continued operability. To be truly fire safe, a valve must be operable even if it is fire-damaged. The best overall design is one that eliminates heat distortion of the valve body and operating mechanism caused by thermal stresses and associated piping stresses during a fire. Some increase in torque should be expected, and so actuators when se-

lected should be sized using an adequate safety factor for worst-case operability.

Types of valves used in fire-safe service

Generally speaking, a fire-safe valve is one that will withstand a fire and provide a degree of shutoff that is acceptable under given conditions. Extremely high temperatures necessitate the use of metal construction. For this reason the first types of valves to be considered fire safe were gate and globe valves because of their metal-to-metal seating. Because of their metal seating these valves leak somewhat in normal operation and will likely leak even more if distorted by a fire. There are no established test standards to measure the fire-safe capability of gate or globe valves.

Today, soft-seated fire-safe valves are preferred because they

1. Provide tight shutoff in normal operation as well as during and after a fire
2. Are economical
3. Are easier to automate than gate or globe valves
4. Are designed and manufactured to meet established fire-safe valve standards

Current soft-seated rotary valves include ball valves, high-performance butterfly valves, and some plug valves. In providing bubble-tight shutoff in normal operation as well as fire safety, manufacturers of rotary stem valves are utilizing two types of seating arrangements with their valves.

The first and most common may be referred to as a two-stage seat (see Fig. 4.19). This system relies upon a full burn (or melt) of the resilient seat before metal-to-metal sealing occurs. For a fire-safe ball

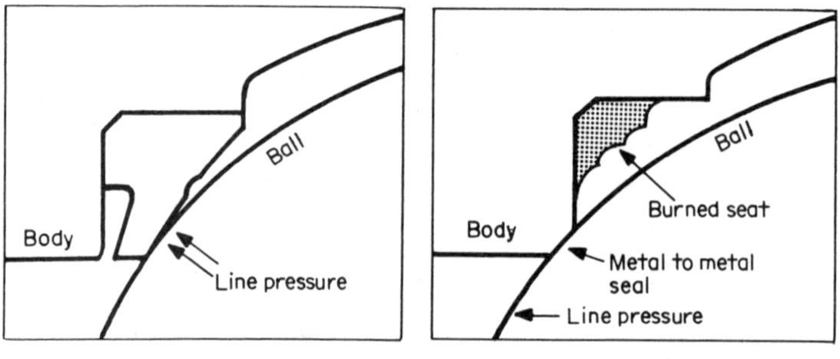

Figure 4.19 Soft seat before fire and metal contact after fire of fire-safe ball valve.

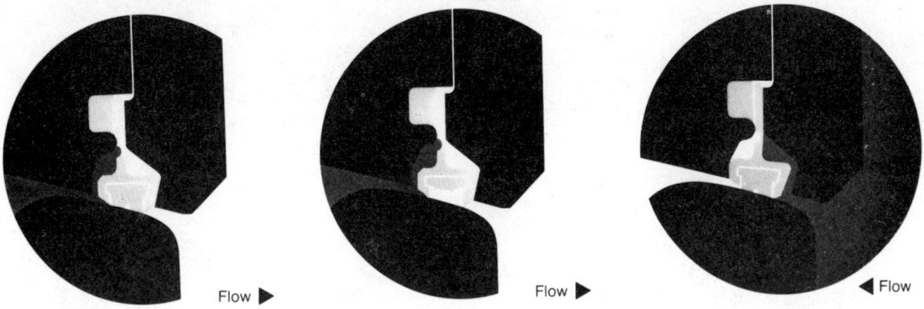

Figure 4.20 Seat design does not require full seat burn due to continuous metal seat contact. (*Courtesy of Xomox Corporation.*)

valve, metal sealing occurs when, for instance, the floating ball moves downstream to contact a machined surface in the body that matches the contour of the ball. If a quickly extinguished fire or other condition prevents full seat burn, however, the floating ball would not be permitted to fully contact the matching downstream metal seat. This could cause excessive internal leakage and defeat the intent of the established testing standards.

A second system typically used by some high-performance butterfly valve manufacturers is not dependent on total seat burn. Here a resilient seat and metal seat contact the disc at the same time (see Fig. 4.20). Because there is always contact with the metal seat it will provide an established leakage rate even if the resilient seat is only partially burned.

External leakage presents other problems. The most common leakage occurs past the valve stem once the thrust washer melts and requires a second metal-to-metal seating arrangement. This is usually accomplished by expanding the outer diameter at the stem's base so that it contacts a lip machined into the valve body. In a manually operated valve, this design works if the handle does not prevent the stem from moving vertically. If the valve is actuated automatically, the drive coupling should accommodate this motion. If the valve is of a two- or three-piece body design, attention must be paid to the body seal materials to prevent leakage during a fire (see Fig. 4.21). High-performance butterfly valves use a rigid disc and stem connection. The packing material is usually graphite based to withstand high temperatures to 1300°F (700°C). The one-piece body design of most high-performance valves eliminates body seal leakage.

Design testing and standards

Since not all fires are alike, safety precautions should not be the same for all situations. While another industry's standards may be followed

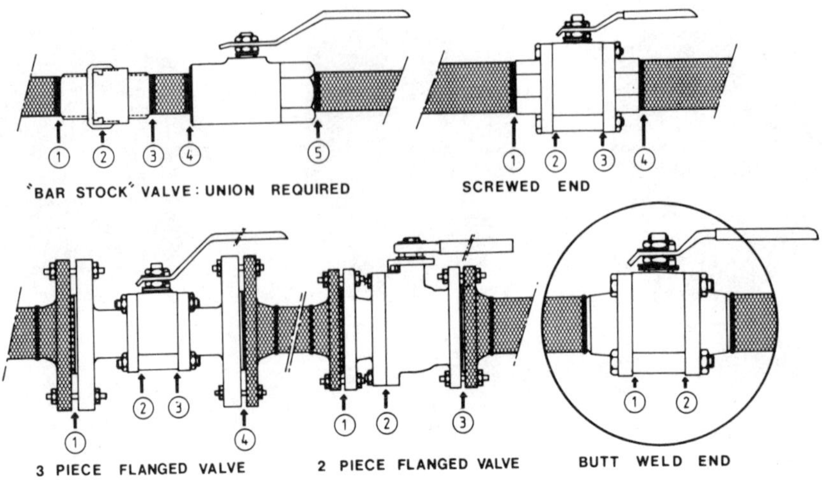

Figure 4.21 Valve body leakage is an important consideration in fire-safe applications. (*Courtesy of MCF Valves.*)

when the medium in the piping is the same, it often is unacceptable to do so when different media are involved. Thus, using standards of the oil-refining industry for other segments used in the chemical process industry raises major questions: Do the refining industry's standards cover the fire hazards posed by media and processes specific to the rest of the chemical process industry? Which criteria come closest to providing proper guidelines for choosing a fire-safe valve for non-oil-refining service?

With the introduction of resilient valve seat materials [rubber or plastic compositions with melting points under 700°F (370°C)], methods of defining and testing the fire safety of soft-seated valves became necessary. In developing a basis for valve design, operating specifications and test procedures, testing houses have first had to specify the fire conditions. Generally, these tests do not duplicate a real fire and so cannot illustrate actual conditions. Because industry experts are unable to agree upon a definition of a standard fire, it is impossible to develop an all-encompassing test for fire-safe valves.

How do testing houses define fire?

Quite often users of fire-safe valves work with in-house or independent test committees to define a fire-safe test to meet their needs. Things to consider for fire-safe valve test specifications are:

1. *Valve type (metal to metal, etc.).* These may be divided into (a) seats which have continuous metal-to-metal contact in the closed po-

sition and (b) two-stage seats which rely upon some secondary means such as seat or ball overtravel, system pressure, gravity, or spring loading to establish metal-to-metal contact when fire occurs.

2. *Stem position.* This is important in evaluating the severity of a fire test and may be the most important criterion in evaluating a test as to its applicability to the chemical process industry. The thermodynamic properties of the medium, vapor pressure, expansion rate, and toxicity must be considered as well. Certainly, a valve with the stem in the vertical position handling a high-vapor-pressure monomer or solvent presents a different problem than one handling a low-vapor-pressure medium such as a diesel fuel. For practical reasons, the more severe vertical stem position should be specified when selecting or defining a test standard.

3. *Bore position.* For purposes of testing, the bore in the horizontal position is often specified so that the weight of the closure element will not augment the seal. This is particularly true for the floating ball valve.

4. *Valve open or shut.* If the valve is open, the test is more severe. While valves in real plant situations are both open and closed, those that are open will be the more critical if, in a fire, they are closed to isolate sections of the plant. In the open position, the valves' soft seat if unsupported can sag into the flow path. Upon closure, the sagging, partly burned seat can prevent full closure of the valve after the fire.

5. *Test pressure during burn.* Each standards organization has made the pressure requirement low. This assumes that most valves are not used at their highest operating pressures and are installed in systems containing pressure-relief devices. The validity of this approach can only be evaluated from the standpoint of good piping practice and safety.

6. *Test medium.* As a test medium, water is safe to use and its properties are easy to measure. However, with water, it may be difficult to spot a leaking packing or body flange without performing a mass balance. In addition, the thermodynamic properties of water and steam may not simulate the real-life fluid. On the other hand, if the test medium is a hydrocarbon, its viscosity and its flammable nature immediately signal a leak during the test. Although this is more dangerous than testing with water, it certainly represents a more realistic situation.

7. *Burn duration.* The time for a test should be based on the specific type and size of valve. During the specified times, a smaller valve would experience total soft-seat destruction. However, a larger one may show only partial destruction. Partial soft-seat destruction is

more realistic during a fire and is a more stringent leakage-test requirement. To accurately and fairly evaluate a fire-safe valve, both partial and total soft-seat destruction should be tested for. The test would encompass the use of a specific time and two temperatures.

8. *Time when seat leakage is measured.* In the Factory Mutual (FM) and American Petroleum Institute (API) tests, seat leakage is measured during and after the fire. Both tests are performed in the closed position. In the Exxon test, the seats are open during the burn. Therefore, the leakage can only be measured after the fire.

9. *Allowable leakage and maximum external leakage.* For the CPI, this is a critical factor, considering toxic leaks, environmental issues, and the like.

Maximum seat leakage. The significance of these leakages can be seen when they are compared with the standards used for valve manufacturers. MSS-SP 61 and API 598 specify 10 cubic centimeters per hour (cm^3/h) and 12 to 20 drops per minute (0.75 to 1.75 cm^3/min), respectively, for new valves.

10. *Operability.* The Oil Companies Material Association (OCMA) test, which requires the valve to open and shut for three cycles when hot, is done within 15 min of the fire test. The valve not only has to be cycled while it is hot but before the leakage testing is done. API requires the valve to be cycled only once after the seat test, therefore not requiring the reseating of the valve. (The testing is done in the closed position.)

Even with their differences, existing tests provide a good general indication of whether a valve is fire safe. For the CPI, users may wish to combine or modify established independent tests to best meet their own needs. Clearly, in setting a standard, one fact should be a primary concern: During a fire, a fire-safe valve will go from having a complete seat to a partially or totally destroyed one. Most industrial fires are quenched long before valve seats are totally destroyed. A partial burn test is the best indicator of how a valve will act during a fire.

Testing criteria and methods. It is impossible to have a single definition of a fire. Certainly the type of fire that may occur in a refinery is different than that in a chemical plant. For example, some fires burn longer or hotter and some spread faster.

The four standards used to establish fire-safe valve performance conditions are

API	American Petroleum Institute
BS	British Standard (formerly the Oil Companies Materials Association [OCMA])

Exxon	Independent Refinery Standards
FM	Factory Mutual Research

These standards reflect what is perceived by each to be important test criteria for the valve sizes and types used in their industry. The differences in flow media, fuel, fire duration, pipe size, and valve orientations, as well as measurement techniques and the amount of leakage deemed acceptable, may be different, but the goal of each test is the same: to establish a minimum safety standard for valves in flammable liquid service.

What is a fire? Before design and performance standards for fire-safe valves are developed, a fire must first be defined by using the following criteria: the fire test medium, the temperature at the valve body, and the duration of the fire. The Factory Mutual Research Corporation requires 15 min of exposure to a flame temperature range from 1400 to 1650°F (760 to 900°C) for the duration of the test. This is considered the heat that is typical of spill fires that might be experienced in a chemical processing plant. To simulate such a fire the test equipment includes a pan of liquid heptane 10 ft^2 (3050 mm^2) in size. The manual or automated valve being tested sits 18 in (460 mm) above the liquid surface for the duration of the burn.

To complete the prefire testing process, the valve must also undergo a cycling test. The valve design and construction must be able to permit reliable operation through 5000 cycles under specified conditions.

After the valve meets normal operating expectations, it is fire-tested. Test equipment includes a pan of liquid heptane 10 ft^2 (3050 mm^2) in size, condenser, safety valve, and measuring device such as a graduated cylinder. Approximately 18 gal (68 L) of heptane, enough to feed a 15-min conflagration, are placed in the pan. The valve, which remains closed throughout the test, sits 18 in (460 mm) above the liquid surface. Its orientation is normal: vertical stem, horizontal bore. Water serves as the flow medium.

During fire testing, the flow media cools only the high-pressure side of the valve; the downstream side suffers the fire's full effect. It takes about 5 min to heat the valve to 700°F (370°C), the temperature at which the soft seat material melts or burns away, triggering the backup seat arrangement. After 15 min, the pan fire is extinguished and the graduated cylinder is removed to measure leakage. The condenser, located downstream of the valve, cools any vapors that might have escaped to keep the reading accurate. The test concludes by spraying the valve with 1¼-in (32-mm) hose for 1 min.

Anything less than 0.057 in^3/min (0.95 mL/min) past the valve seat, regardless of pipe size, is acceptable. External leakage around the stem or flanged fittings is limited to individual drops. The requirements also apply to ordinary valves. If they fail to meet the requirements, testing is suspended temporarily. The manufacturer may then withdraw the unit or implement the design changes needed to meet the requirements.

Fire-safe valve specifications summarized

Regardless of safeguards taken during the design process, a valve design should not be considered fire safe until it has been tested under fire conditions. Standards such as those sponsored by the API, BS, Exxon, and Factory Mutual Research are all comparable. See Fig 4.22 for fire test comparisons. For valves to be approved by a third party, the valve manufacturer must meet certain quality standards during manufacture. A site visit is usually conducted to establish the control conditions under which the valve is produced, and to determine the manufacturer's commitment to product quality.

In a multimillion-dollar chemical process, valves play a role disproportionate to their size and cost. These small, relatively inexpensive parts are critical to the safe performance of the process. Since valves are often the first line of control for flammable liquids, it is extremely important to install only valves that are fire safe. This one preventive act can help save hundreds of thousands of dollars and hinder costly business interruptions.

Test Specification	BS5146 and OCMA FSV.1	Exxon BP3-14-1	API 607 Third Edition	FMRC 7440	API RP 6F First Edition
Stem position	Vertical	Vertical	Horizontal	Not specified	Horizontal
Bore position	Horizontal	Horizontal	Horizontal	Not specified	Horizontal
Valve open or shut	Open	Open	Shut	Shut	Shut
Test pressure during burn	30 psi	25 psi	75% of cold working pressure	125 psi	Depends on valve rating
Test media	Kerosene or diesel fuel	Liquid hydrocarbon	Water	Not specified	Water
Valve body temperature	Sufficient to destroy soft seat	1200°F minimum	1400°F minimum	Not specified	Not specified
Burn duration	15 min.	15 min.	30 min.	15 min.	15 min. or 30 min.
When seat leakage measured	After test	After test	During test and after test	During test	After test
Maximum external leakage	No appreçiable leakage	Leakage shall be negligible	20 ml/min/in. diameter	0.1 qt/min (94.6 cc/min)	100 cc/min per inch of valve size
Maximum seat leakage	10 ml/min/in. diameter	10 ml/min/in. diameter	40/ml/min/in. diameter	Individual drops	400 cc/min per inch of valve size
Operability	3 cycles open to shut	3 cycles open to shut	1 cycle open to shut	Must be operable	Shut to open against test pressure differential

Figure 4.22 Comparison of fire test specifications. (*Courtesy of Valve Manufacturers Association of America.*)

Part

2

Actuators

Chapter

5

A Review of Manual Valve Operator Designs

5.1 What Is a Valve Actuator?

An actuator may be defined as a power device that produces torque (or thrust) to move the stem (and thus the closure member) of a valve for flow control. Actuators are available as pneumatic, hydraulic, electric, or electrohydraulic power units and may be required to cycle a valve in a millisecond or very slowly—up to 20 min or slower depending on the application. They can provide fail-safe action, and with the use of certain control devices they can "lock up" a valve in the event of some catastrophic system failure.

In throttling applications actuators must respond to a control signal and precisely position the valve to regulate flow. Actuators must certainly be able to withstand harsh environments with extreme heat and cold, but they must also withstand corrosion. Control accessories (positioners, solenoid valves, limit switches, etc.) for any actuator must comply with the control system requirements and must also meet the area classification [hazardous (NEMA VII), weather resistant (NEMA IV), etc.].

Most valves are operated manually, by causing some linear or rotational movement through a handwheel, lever, wrench, gear, etc. Because of the variety of applications, locations, and flow control functions quarter-turn valves serve, there is a correspondingly wide selection of manual operators in a variety of configurations available. Manually operated valves are usually designed for reasonable required physical effort applied to a handwheel or handle and are intended to assure that valve operation is in the correct direction, is not too fast or too slow, and is applied through the proper distance. Properly designed handles may also serve as valve position indicators. When a handle is positioned parallel to the pipeline, this usually indicates that the valve is open. Valve manufacturers select gearbox ratios and/or handle lengths to minimize the manual effort. The basic manual operator is the two-position lever or handle. It is designed to move the valve to the open or closed position. Lever operators can be used for control of butterfly valves that are 5 in (12 mm) and smaller or up to 12 in (300 mm) in size, if flow rates are less than 5 ft/s (1.5 m/s). Properly applied, levers provide quick valve action, economy, and simplicity. A variation on that is a 10-position lever lock handle available for butterfly valves and some ball valves. The spring-loaded trigger fits into one of 10 different notches to allow intermediate positioning of the valve disc in 10 percent increments. This is generally not necessary for ball or plug valves because they are inherently stable and will remain in an intermediate position without the need for locks. This also means that they are also infinitely adjustable for precise flow control. Butterfly valves have optional handles that provide for infinite position adjustment. Two options available with lever operators are rate-of-flow memory stop devices and spring-return handles. If the valve is to be closed for any reason, the rate-of-flow memory device allows the valve to return to its original regulated position to regain the previous rate of flow in the system. In a spring-return handle there is a spring in the handle which ensures that when the valve is not in manual control, the valve always returns to a predetermined position—most often closed. Depending on the service, one should determine if a fully enclosed spring is necessary to eliminate spring corrosion.

Other variations on the lever are oval handles, stem nuts, T-bars, and chain levers (see Fig. 5.1). Oval handles are used on small ball valves when they are located in a service where an accidental bump into a lever may be disastrous. The oval handle also looks more attractive in panel-mounted installations. A stem nut is usually used in underground service. A wrench is used to open and close the valves. Because the wrench must reach below ground, the handle of the wrench is usually a double lever (or a T-bar) to prevent side loading. Chain levers are used for overhead installations. The ends of a single chain are connected to a T-bar on the valve or to a chainwheel on multiturn valves.

Throttling valves 8 in (200 mm) or larger should have an infinite position-throttling operator, such as a gearbox and handwheel. This gives the operator better control against the possibility of the valve closure element slamming against the seat when severely throttled, which might result in system damage. This type of operator also allows greater ease of operation and closer settings. For throttling valves, lever lock handles are not satisfactory because of their relatively few position set-

Figure 5.1 Selection of rotary valve manual operators and options. (*Courtesy of Velan Inc.*)

tings, usually 10 or 12. In HVAC applications, very close settings are needed to precisely regulate and properly balance the system.

5.2 Fluid Dynamics of Manual Valves

A flowing fluid in a piping system has mass and velocity. Anything that causes a moving mass to change its velocity will experience a reacting inertia force in proportion to the magnitude of the mass and the rate of the imposed velocity change. When the closure member of a valve is subjected to a flowing fluid, these reactions are manifested in the form of valve torque. Positive torque would increase the required effort to operate a valve. Manual as well as automatic operators must be sized to overcome this dynamic torque. For our purposes negative torque may be considered the result of fluid forces assisting the rotation of the closure element. Negative torque is manifested in a butterfly valve when it tends to close itself during manual operation as lift and drag forces are exerted on the disc. This represents a potential safety and system problem. To minimize the danger, it is often recommended that butterfly valves have gear operators when they are used in high-velocity service. Another advantage of a gear operator is that it provides precise incremental positioning of a valve to regulate flow.

Chapter

6

Pneumatic Actuators

6.1 Fundamentals of Power Actuator Selection

As the process industry continues to achieve more efficient and productive plant design, plant engineers and technicians are faced, almost daily, with new equipment designs and applications. One product, a valve actuator, may be described by some as simply a *black box,* having an input (power supply or signal), an output (torque), and a mechanism or circuitry to operate a valve. Those who select control valves will quickly see that a variety of valve actuators are available to meet most individual or plantwide valve automation requirements. In order to make the best technical and economical choice, an engineer must know the factors that are most important for the selection of actuators for plantwide valve automation. Where the quality of a valve depends on the mechanical design, the metallurgy, and the machining, its performance in the control loop is often dictated by the actuator.

The decision to automate a valve is usually based on one or all of the following considerations.

1. Safety
2. Reliable operation
3. Control and process system performance
4. Inaccessible or remote valve location
5. Cost
6. Excessive valve torque
7. Emergency response and whether it is fail-safe

All actuators have several distinct purposes. They must

1. *Move* the valve closure member (disc, ball, or plug) to the desired position. Not only must the actuator provide enough torque or thrust to move the closure member under the most severe conditions, it must also be fitted with the appropriate controls to direct it.

2. *Hold* the valve closure member in the desired position. Particularly in throttling applications where fluids may create a dynamic torque, actuators should have adequate spring or fluid power or mechanical stiffness to overcome this phenomenon.

3. *Seat* the valve closure member with sufficient torque to provide the desired shutoff specification. A butterfly valve for instance is fully seated (closed) when the disc is positioned in a resilient liner (seat). In this rotary position the valve stem torque is at its highest. Actuator sizing for torque-seated butterfly valves may require special accessories particularly on electric actuators to ensure that sufficient torque is sustained in the closed position.

4. Provide a *failure mode in the event of system failure.* This may be fully opened, closed, or as-is depending upon the application. Certain failure mode requirements may eliminate electric actuators yet be ideal for pneumatic or electrohydraulic units.

5. Provide the required rotational *travel (90°, 180°, etc.).* Valves requiring more than 90° of rotation include multiported valves. A few pneumatic actuator manufacturers offer 180° actuators. For greater than 180°, electric actuators are usually preferred because they are electrically, not mechanically, limited in rotation.

6. Provide the required *operating speed.* All actuators may be regulated in cycle speed depending on the control circuit elements used.

Fast cycle speeds (less than one-half the standard actuator cycle time) require careful valve selection. The physical shocks associated with fast cycling can damage the valve parts—especially when combined with high cycle rates. Special preparation of pneumatic actuators—including special solenoids, piping, and quick-exhaust valves—may be required to achieve high cycle speeds.

The cycle speeds of electric actuators cannot be increased, only slowed. This is easily accomplished with the specification of either special cycle times or with the addition of an electronic speed control card. Special cycle times are achieved with a different gearing mechanism which also affects output torque. The electronic speed control is infinitely adjustable and can reduce the effective actuator speed up to 20 times without the need for special gearing. Output torque of the actuator is not affected where speed cards are used. Pneumatic actuators can be slowed by the use of speed control valves in the air piping. One speed control valve will slow speed in one direction, while two are required to slow speed in both directions. Speed controls do not affect the output torque of pneumatic actuators.

High cycle rates will require special selection and preparation of valves and actuators. A high cycle rate for pneumatic actuators is defined as cycling continuously in excess of 30 times per hour; electric actuators used in excess of a 25 percent duty cycle are said to have a high cycle rate. High cycle rates place additional stress and wear on the valve stem. The small amount of play between the actuator, stem, and ball increases the wear in the stem in high cycle rate applications. This is minimized with special preparation to assure long life of the valve package.

The actuator specification form (Fig. 6.1) may prove helpful when gathering pertinent actuator selection information.

Pneumatic and electric actuators compared

At times it is necessary for a process engineer to choose between a pneumatically or electrically actuated valve for a system. There are advantages to both styles, and it is valuable to have data available to make the best choice.

Compatibility (power source). First and foremost in the selection of an actuator type (pneumatic or electric) is to determine the most effective power source for the actuator. Points to consider are

1. Power source availability
2. Torque at valve stem
3. Failure mode
4. Control accessories
5. Speed of operation
6. Frequency of operation
7. Plant environment
8. Size of valve
9. System component costs
10. System maintenance

The most practical pneumatic actuators utilize an air pressure supply of 40 to 120 lb/in^2 (3 to 8 bar). Generally they are sized for a supply pressure of 60 to 80 lb/in^2 (4 to 6 bar). Higher air pressure is usually

Data Form for Specifying Automated Valves

Section I: **Pneumatic Actuators and Controls for On/Off and Throttling Control**
Section II: **Electric Actuators and Controls for On/Off and Throttling Control**

To insure automated valve performance in todays sophisticated process control systems it is essential to develop a complete specification for each of the elements in the control package.

Though the valve actuator is the basic element in a valve automation package, pilot valves, limit switches, positioners, speed controls, transducers and other control accessories are necessary to properly communicate with the system and to characterize the performance of the automated valve to meet the process/flow control requirements.

The operating environment must also be considered and measures taken to insure compatibility. This may include specifying the materials of construction or electrical design for area classification.

This data form is intended as a specifying aid for valve, actuator and control system elements and to focus upon the design and performance characteristics of the most common automated valve components.

The following guide is for actuators and controls only. It is assumed that the valve to be automated has already been selected based on its own unique ability to handle the service for which it is intended. A complete valve specification for which an automated package may be specified should include:

-Valve size ______Type ______Mfr.______ Model Nr.______ Valve Torque ______ (if available)

-Service Conditions:
Line pressure ________ Media Temp. ________°F Media (Slurry, Gas, etc.)________

Actuator Sizing

The most important step in developing an automated valve specification is to determine a sizing criteria. If a valve is to operate in a process handling clean liquids at moderate pressures and temperatures, the manufacturers published operating torque is usually adequate for actuator sizing. Under certain conditions, however, the torque required to operate a valve may increase. In this case a sizing safety factor may need to be applied based on the following guidelines:[1]

Dry Gas/Air	-	25%(50% for rubber seated valves)
Slurry	-	25-50%
High Velocity	-	25%
High Pressure	-	25%
High Temperature	-	25%
Infrequent Cycling[2]	-	25%
Retro-fitting[3]	-	25%

[1] Consult the valve manufacturer for specific Safety Factor recommendations.
[2] A valve that sits in one position for long periods may require more operating torque. Consult valve manufacturer.
[3] Manual valves to be converted to automated valves should be checked for operating torque.

Figure 6.1 Actuator data form.

difficult to guarantee, and lower pressures require a very large diameter piston or diaphragm to generate usable torque.

Electric actuators are often used with a 110-V ac power supply but are available with a wide variety of ac and dc motors in single phase and three phase.

Temperature range. Both pneumatic and electric actuators may be used in a wide temperature range. The standard temperature range of

Section I: Pneumatic Actuator Specification Data

A. Sizing Data

-Valve Torque: _______ lbf.in.
-Sizing Safety Factor: _______ %
-Double Acting or Spring Return: _______ .
-Fail Open: _______ Fail Closed: _______
-Fail on: Electric only _______ Pneumatic only _______ Electric or Pneumatic _______
-Operating Temperature Range: _______ to _______ °F
-Min/Max Operating Speed (if critical) _______ sec/90° (open) _______ sec/90° (close) (Also important to consider in pneumatic actuator sizing is the minimum air pressure available to power the actuator. If a range of pressure is available (60 - 90 psig for example) the minimum should always be specified for actuator sizing).
-Minimum air pressure ___________ psig.
-Materials of Construction/Environmental Specs _______ (i.e., caustic wash, aggressive fumes, etc.)

B. Pilot Valve

Pneumatic actuators are usually piloted by a solenoid valve. The actuator manufacturer will size the valve flow rate to provide adequate speed of operation and economy. If faster or slower speeds are critical to a process this must be specified for more accurate pilot valve sizing. The following valve types are generally used for piloting actuators:
-Double Acting Actuator 4 way, 5 ported, single coil
-Spring Return Actuator 3 way, 3 ported, single coil

Except where noted the following specification guideline applies to both 3 way and 4 way valves:
-Remote, local or direct mount: _________
Remote: Located away from the actuator in a control panel or on a manifold.
Local: Located at the actuator and adapted to the supply port of the actuator with tubing and fittings.
Direct: Some actuator manufacturers offer an integral air manifold to which a solenoid valve may be direct mounted to eliminate tubing and fittings.
- Materials of construction (aluminum, bronze, stainless steel) _________ (aluminum is standard)

Coil Specifications -
- Coil voltage: _________ AC/DC (standard: 110, 24, 12 VAC/DC)
- Class A (230 degrees F): _________ Class F (310 degrees F): _________
- Class H (350 degrees F): _________
- Area classification: (NEMA IV [weatherproof], VII [hazardous], etc) _________
- Energize to (open or close) _________
(When the pilot valve is energized will it drive the process valve open or closed?)

Pilot Valve Options -
- Manual override: Maintained _________ Momentary _________ Lock _________
- Air Pilot: yes _________ no _________ (pilot valve action initiated by air signal)
- Single or Dual Coil : _________ (Dual coil available on 4 way valves only)

Speed control -
The speed of operation of a pneumatic actuator can be regulated in one or both directions of a cycle. Most pilot valves are designed with NPT exhaust ports that may be fitted with a needle valve or speed control valves.
- Specify: Open direction ________ seconds Close direction ________ seconds

Figure 6.1 *(Continued)*

a pneumatic actuator is from −4 to 175°F (−20 to 80°C) but may be extended to −40 to 250°F (−40 to 121°C) with optional seals and bearings. If control accessories are used (limit switches, solenoid valves, etc.), they may not have the same temperature rating as the actuator, and this should be considered in all applications. In low-temperature applications, the quality of the supply air pressure in relation to dew point should be considered. Dew point is the temperature at which

C. Limit Switches

Limit switches are used with pneumatic actuators to indicate the open, closed, or intermediate position of the process valve. As an electrical device limit switches are enclosed in a housing designed to meet area classifications.

-NEMA Rating: NEMA 4 _________, NEMA 7 _________, NEMA 4-X _________
-Switch styles: SPDT _________, DPDT_________, SPST_________, etc.
-No. of switches: Open _________ Close _________ other _________
-Electrical Rating: (expected current draw, voltage) _________ amps _________ Volts
-Visual Indicator: yes _________ no _________
-Switch designs: Snap acting
Hermetically sealed reed switches
Non-contact (proximity type)
Others--consult actuator supplier

D. Positioners

Positioners are used with piston actuators and some diaphragm actuators to convert a control signal to a valve position in modulating service. There are two basic styles of positioner: pneumatic and electro-pneumatic.

1. Pneumatic Positioners
- Input control signal: 3 - 15 psig etc. _________ psig
- Signal/position: valve open _________ psig valve closed _________ psig
- Cam Style: * linear _________ quick opening _________ other _________
- Visual Indicator: yes _________ no _________.

2. Electro-Pneumatic Positioners
- Input control signal: 4 - 20 maDC etc. _________
- Signal/position: valve open _________ ma
valve closed _________ ma
- Cam Style: * linear _________ quick opening _________ other _________
- Agency approvals (if necessary): (FM, etc.) _Area Classification:_________
- Options: Gauge set
Internal limit switches _________

*Most off-the-shelf quarter turn valves have equal percentage flow characteristics. This is generally acceptable for most applications. If other characteristics are required a replaceable characterized cam may be used with most positioners.

3. Rotary Position Transmitter

A position transmitter provides continuous feedback of the valve position with a resistive or current output signal.
-Resistive Feedback is achieved through a potentiometer but may not be accurate over long distances due to the inherent resistance of the wire.
- Specify resistive range/90° of rotation: zero to_____ohms.
-Current Feedback is achieved through a resistance-to-current conversion circuit (RI) and is generally preferred for transmission over long distances. Output is 4-20 maDC over 90° of rotation.
-NEMA 4 _________ NEMA 7 _________

Figure 6.1 *(Continued)*

condensation occurs in air. Condensate may freeze and block air-supply lines, making the actuator inoperable.

Electric actuators are available in a temperature range of −40 to 150°F (−40 to 65°C). When used outdoors, an electric actuator should be sealed from the environment to prevent the introduction of moisture to the internal workings. Condensation may still form inside, if drawn from the power-supply conduit, which may have captured rainwater prior to installation. Also, since the motor warms the inside of

Section II: Electric Actuator Specification

Electric rotary actuators are basically geared motors with a variety of electrical accessories (limit switches, heater, circuit boards) that may be specified to meet the exact valve automation requirements of a process control system. As an electrical device it is housed in an enclosure designed for specific area classifications (NEMA Classification).

A. Specification Checklist:

- Weatherproof (NEMA 4): _______ Hazardous (NEMA 7)[1]: _______
- Motor Voltage: 12 _____ 24 _____ 110 _____ 220 _____ VAC or VDC
  3-phase _____ specify voltage [2] _____
- Compartment heater: yes _____ no _____

[1] Specify: Class _______ Division _______ Groups _______
[2] 3-phase motors require a reversing starter and/or transformer that may be located:
Remote _____ or integral _____ Specify Control Voltage _________VAC

B. Torque Switches

Torque switches are recommended on electric actuators with an output torque of 800 inch pounds and higher to protect the automated valve and motor. They are also necessary for tight closure of torque seated valves.

- Torque switch: Open yes _____ no _____
  Close yes _____ no _____

C. Manual Override

A manual override is designed to provide manual valve operation in the event of electrical failure or for start-up procedures. The handwheel should not rotate during electrical operation.

- Manual override yes_____ no _____

D. Separate Controls

- Single Phase motor:
  Local/Remote Switch _______
  3 Position Rotary Switch _______
  (open/stop/close) _______

- 3-Phase Motor
  Local/Remote Switch _______
  3 Pushbuttons
  (open/stop/close) _______

E. Electronic Speed Control

As a geared motor an electric actuator has a fixed speed of output rotation that is a function of the layout of the gear train. Some manufacturers offer their products with a rotary output speed of 2 seconds but most are in the range of 5 to 25 seconds for 90° depending upon the model. If a cycle time greater than what is available is necessary an electronic speed control should be specified.

- Electronic speed control: yes _____ no _____ Speed of operation _______ sec/90°min.

Figure 6.1 *(Continued)*

the actuator enclosure when it is operating and cools it when it is not, temperature fluctuations may cause environmental “breathing” and condensation. For this reason, all electric actuators used outdoors should have a heater installed to maintain a constant temperature and eliminate breathing.

It is sometimes difficult to justify the use of electric actuators in a hazardous environment, but if compressed air is not available or if a pneumatic actuator will not provide the operating characteristics required, then an electric actuator with a properly classified enclosure may be used.

F. Position Transmitter

A position transmitter provides continuous feedback of valve position.
- Slide wire (pot) only _______ Resistance _______ ohms/90°
- 2-wire 4-20 maDC output _______ (integral power supply)
- 4-wire 4-20 maDC output _______ (requires a power supply)

G. Limit Switches

Limit switches are used to provide the necessary number of electrical contacts in the open, close or intermediate position. Standard electric actuators have two SPDT limit switches, one for the open position and one for the closed position to control rotary travel as desired.

Specify the number of additional contacts required in each position:
- Open _______ contacts (make)
  _______ contacts (break)
- Close _______ contacts (make)
  _______ contacts (break)

-Intermediate position:
Specify rotary position in degrees _______ °
_______ contacts (make)
_______ contacts (break)

H. Positioners

Positioners are used with electric actuators to position the valve in response to a control signal.
-Position transmitter required: yes_______ no_______ Input signal: _______ maDC;VDC or ohms

I. Control Options for Electric Actuators

One of the most appealing characteristics of electric rotary actuators is its inherent versatility. Electronic control cards, switches and wiring are easily re-configured to meet the control system needs.
-3-position control package provides 0-90-180 rotation for multi ported, 3 position valves.
-3-position options: yes_____ no_____
-mid position stop from both / either extreme. If either, then from: CW_____ CCW _____
-Integral Local Control Station - mounts directly to actuator housing and includes Local / Remote switch plus open/close switch for local operation.
-Local Control Station: yes _____ no_____
-Options: Key lock _____
Open/Close lights _____

J. Other Electric Actuator Options

-Interposing Relay: _________.
-Plug & Socket Connector: _________.
-DC Motor / Battery Backup for Fail-Safe Operation: ________.
-Special Coatings: _________ (Specify environmental conditions)

Figure 6.1 *(Continued)*

NEMA guidelines

The National Electrical Manufacturers Association (NEMA) has set up guidelines for the construction and installation of electric actuators (and other electric devices) for use in hazardous environments. The NEMA VII guidelines read

> VII Hazardous Location Class I (Explosive Gas or Vapor)
>
> Meets application requirements of National Electrical Code; conforms with specifications of Underwriters' Laboratories, Inc., used for atmosphere containing gasoline, hexane, naphtha, benzine, butane, propane, acetone, benzol, lacquer-solvent vapors, and natural gas.

Almost all electric actuator manufacturers have an option for a version of their standard product line that conforms with NEMA VII.

On the other hand, pneumatic actuators are inherently explosion-proof. When electric controls are used with pneumatic actuators in hazardous areas, they are generally more cost-effective than electric actuators. Solenoid-operated pilot valves may be mounted and powered in a nonhazardous area and piped to the actuator. Limit switches, for position indication, may be housed in a NEMA VII enclosure. The inherent safety of pneumatic actuators in hazardous areas makes them a practical choice in these applications.

Spring return. Another safety accessory widely specified in the chemical process industry on valve actuators is the spring-return (fail-safe) option (see Fig. 6.2). Upon power or signal failure, a spring-return actuator drives the valve to a predetermined safe position. This is a practical and inexpensive option with pneumatic actuators and is an important reason for the wide use of pneumatic actuators throughout the industry.

Where springs are not practical because of actuator size or weight, or if a double-acting unit is already installed, an accumulator tank may be installed to store air pressure. Electric actuators are not widely available in a spring-return version. There is at least one manufacturer who has a unit that is neat and compact, but the limited torque [600 lb · in (68 N · m)] makes it impractical for plantwide use. To accomplish the spring-return function, however, an electro-hydraulic actuator is often a good choice. Electrohydraulic actuation is achieved by energizing a hydraulic pump, which pressurizes a single-acting (spring-return) cylinder. Upon power failure, spring action drives the actuator to its original position. Because only an electric power supply is required for this self-contained unit, it is a practical approach to fail-safe electric operation.

Performance characteristics. Before specifying a pneumatic or electric actuator for valve automation, it is important to consider a few of the key performance characteristics of each.

Duty cycle. Pneumatic actuators have a 100 percent duty cycle. In fact, the harder they work, the better they work.

Electric actuators are most commonly available with 25 percent duty cycle motors. This means that to prevent overheating the motor the unit must be turned off for three cycle periods for each operating cycle within a 15-min period before being operated again. Because most on-off automated valves remain in one position for as much as 95 percent of the time this is not usually a problem. Accurate sizing is

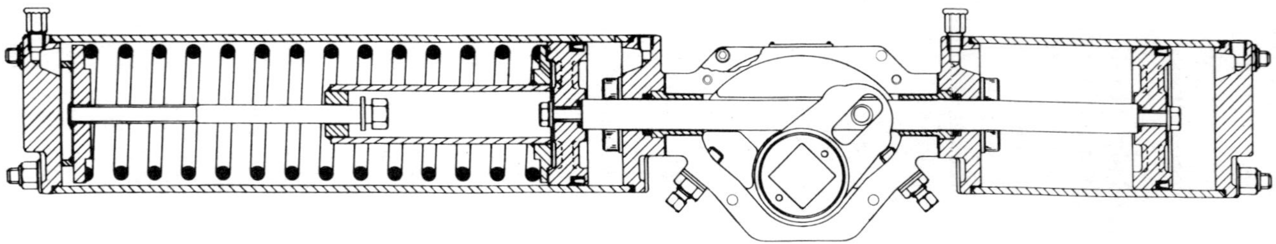

Figure 6.2 View of spring-return pneumatic actuator. (*Courtesy of Posi-seal International, Inc.*)

important however. With optional motors and/or capacitors, an electric actuator may be upgraded to 100 percent duty cycle.

Stalling. Pneumatic actuators may be stalled indefinitely without overheating. In some instances, if determined to be safe, the supply air pressure may be increased within the limits set by the manufacturer to overcome temporary high-torque requirements.

Electric actuators should not be stalled. Stalling an electric motor draws excessive current, which generates heat in the motor and can cause damage. These actuators may be fitted with a heat-sensing element that is embedded in the windings of the motor and in series with the power wiring. At a specified temperature, the sensor opens the power circuit to the motor; when cooled, the circuit closes again.

Another device, a torque switch, is designed to sense the power being transmitted through the actuator gearing. As the torque required by the valve approaches the limit of the actuator, the torque switch opens the power circuit to avoid motor, gear, and valve damage. The mechanism is available in a variety of designs, and some are adjustable to trip at a desired torque value.

Speed control. The ability to control the speed of a pneumatic actuator is one of the most underutilized advantages of the design. Speeds from 2 to 20 min are easily achieved with many pneumatic rotary actuators. The benefits of this are prevention of water hammer, controlled steam injection speeds, and sequencing, to name a few.

The simplest way to control the speed of a pneumatic actuator is to install a variable orifice on the exhaust port of the actuator (or solenoid valve). By adding a few simple components to the assembly, two-speed operation may even be achieved.

Increasing the speed of a pneumatic actuator. Pilot valves used with pneumatic actuators may be specified with higher flow coefficients C_v through both the supply and exhaust ports to increase the speed of operation. Attention must be paid to any internal porting within the actuator to ensure that the higher flow rate through the pilot valve is not restricted once it enters the actuator.

Speed control for an electric actuator. Since an electric actuator is basically a geared motor, it is impossible to make it cycle faster unless a gearing change is made. For slower operation, a pulsing circuit, which uses an electronic timer with a setting for the *on* period and the *off* period, can be added. When energized, the timer will power the motor for a given period of time and then shut off for another. This alternating on and off pulsing continues until the actuator reaches its limit of rotation. As an example, if an actuator cycles 90° in 15 s and a pulse

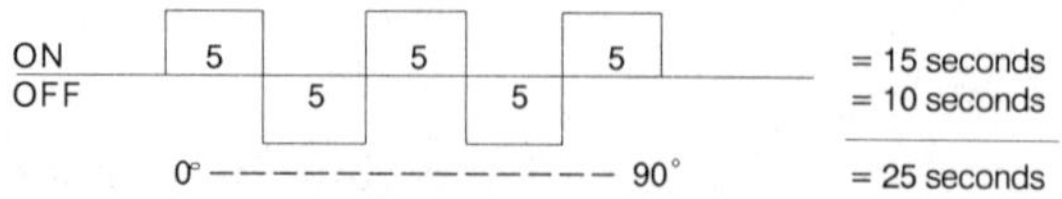

Figure 6.3 Example of how a pulsing circuit extends the operating time of an electric actuator.

circuit is installed and set to 5 s *on* and 5 s *off,* it will take 25 s to cycle 90° (see Fig. 6.3).

Modulating. In modulating service, an electric actuator interfaces well with existing electronic control systems and eliminates the need for electropneumatic controls. For pneumatic actuators, there is a variety of positioners available for modulating service. These devices accept a pneumatic control signal but are also available with transducers to convert electronic signals to useful pneumatic signals.

Torque-to-weight ratio. Electric actuators have a high torque-to-weight ratio above 4000 lbf · in (450 N · m), and they easily adjust, with limit switches, to virtually any intermediate position. Pneumatic actuators have an excellent power-to-weight ratio, up to about 4000 lb · in (450 N · m).

Summary of pneumatic and electric actuators

This table of characteristics summarizes the comparison of pneumatic and electric actuators.

Pneumatic	Electric
Simple, accurate, and inexpensive speed control	A pulsing circuit may be added to slow the operating speed
Inherently explosion-proof, spark-proof	Available with a NEMA VII enclosure for hazardous areas
Not subject to overheating; not sensitive to wet environment	Motor designed to prevent current or temperature damage. Must be sealed from moisture; heater and thermostat required
100% duty cycle	25% standard duty cycle. May be upgraded
May be stalled indefinitely	Should not be stalled
Torque-to-weight ratio averages 123:1 at 1500 lbf · in (170 N · m)	Torque-to-weight ratio averages 44:1 at 1500 lbf · in (170 N · m)
Spring-return (fail-safe) option is practical and economical	Electrohydraulic actuator is a good choice for electric fail-safe function

6.2 Understanding Valve Operating Characteristics

Whether a valve is linear (multiturn) or rotary (part-turn), the task of selecting and specifying the automation package must include a review of valve operating characteristics. Difficult or expensive actuator parameters for the specific type, size, or service of a valve may require that it be replaced or that other means of automation be considered.

Operating characteristics—linear valve

Gate, globe, pinch, and diaphragm valves are considered linear (multiturn) valves because the linear action of the valve stem provides the action necessary to effect the required operation—open, close, or set in between. When manually operated, these valves usually have a handwheel that when rotated acts upon the threaded stem to raise and lower it. When the handwheel is removed for actuator mounting, the reciprocating (up-down) action may be performed by a pneumatic, hydraulic, or electrohydraulic actuator or the multiturn action may be performed by a threaded drive nut driven by an electric actuator (see Fig. 6.4). The purposes of an actuator are to overcome the friction,

Figure 6.4 Electric actuator mounted to multiturn gate valve. (*Courtesy of Rotork Controls, Inc.*)

pressure, and other forces acting on the valve to perform automatic operation; to hold the valve in the desired position under extreme conditions; and to provide fail-safe action, to name a few.

To ensure that these forces may be overcome it is necessary to accurately size an actuator. The best way to size a linear valve is to quantify the operating forces which are based on valve size, operating pressure, style, or materials. The opening and closing of a linear valve requires thrust. Based on unique factors for calculating thrust, linear valves will have operating forces that are characteristic of their design. Once determined, actuator selection and sizing may begin.

Valve thrust calculation. Determining the operating thrust of a gate or globe valve should be done with the assistance of the valve manufacturer. The methods often require conservative estimates along with other specific information such as

1. Valve type
2. Valve size
3. Maximum differential pressure (the inlet pressure less the outlet pressure), which usually occurs in the closed position for on-off valves
4. Valve stem diameter (Each actuator has a maximum allowable stem diameter.)

This information is necessary to calculate the amount of thrust that a linear actuator must produce to achieve the opening and closing of the valve. Once acquired, the information is used for the following calculations.

$$\text{Gate valve thrust} = (\text{valve size})^2 \times \text{differential pressure} \times \text{valve factor} \quad (\text{VF} = 0.3)$$

$$\text{Globe valve thrust} = \text{port area} \times \text{differential pressure} \times \text{valve factor} \quad (\text{VF} = 1.25)$$

From this it can be determined that with a 150-lb/in^2 differential pressure a 4-in gate valve will require the following operating thrust:

$$(4)^2 \times 150 \times 0.3 = 720 \text{ lb}$$

and a 4-in globe valve with a 3-in port area will require

$$\pi(1.5)^2 \times 150 \times 1.25 = 1325 \text{ lb}$$

Valve factors are used to include the effects on the total thrust required by stuffing box friction and the weight of the gate.

Operating characteristics—rotary valves

The most popular type of valve used throughout the CPI is the rotary valve. This includes ball, plug, and butterfly valves in both quarter-turn rotation (90°) and extended-rotation (180°) valves such as multi-ported valves. The closure element of a rotary valve is controlled by a valve stem to which a handle, gear, or actuator may be fitted.

Understanding valve torque. For a rotary valve, torque represents the amount of turning force required to cycle from open or close under process conditions. To operate a valve with an actuator, the torque value of the actuator at a specified air pressure should exceed the rated valve torque by at least 10 percent. This is known as a sizing safety factor, and it may vary by manufacturer or service.

The valve manufacturer usually publishes a single torque value for a given size valve, which is the maximum value at any point in the cycle of the valve but is usually the breakaway (unseating) position. Torque is determined by several factors including

1. Line pressure
2. Stem-packing tightness
3. Set period
4. Temperature
5. Viscous or slurry media
6. Age
7. Fluid velocity

For example, a ball valve has a maximum torque (the manufacturer's published value) and a minimum torque. The maximum torque usually occurs at the breakaway point, or when the valve begins to move from the closed position. From there, the torque required at any point in the cycle will remain less than the published torque but can vary because torque is affected by factors previously detailed.

Before the actuator can be sized for any given valve application, the amount of torque required by the valve must be determined. The operating torque of the ball valve is influenced by a number of factors. Some are related to design and material; others are related to the application (service conditions). Design-related factors include the type and material of the valve seats. Application factors include system pressure, media, and frequency of operation.

The torque required to operate a ball valve comes from two different areas within the valve, in both cases resulting from friction between metal and relatively soft sealing materials. These two areas of contact in the valve that create torque are the stem and packing and the ball and seat.

Stem torque is primarily dependent upon the tightness of the stem nut which serves to compress the packing against the stem. Proper adjustment of the stem nut is important to valve performance and life. If the nut is too loose, the valve exhibits stem leakage; if the nut is too tight, the total torque requirement can be increased to the point where the actuator may not be strong enough to cycle the valve.

Ball and seat torque is created by the interference between the ball and seat and is very sensitive to service conditions. The *floating ball* design concept allows the system pressure to force the ball into the downstream seat. The higher the system pressure, the harder the ball is forced into the seat, and therefore the higher the torque. Since different seat materials have different coefficients of friction, the ball and seat torque also becomes a function of the seat material being used.

Valve torque is also affected by the media flowing through the valve. Abrasive media have a tendency to increase the amount of torque required, whereas some light oils, which provide additional lubricity, can reduce the amount of torque required. When valves are used in gas service, the seat and ball contact area is extremely dry and will exhibit higher friction and require higher operating torque.

The most often overlooked factor affecting valve torque is frequency of operation. When a valve is cycled regularly, the torque requirement tends to decrease (to a point) with time. However, when the valve sits in the closed position for an extended time, the torque requirement will increase almost to the point where it was when the valve was new. The torque increases because the soft seat material tends to flow into the surface imperfections of the closure element, giving a degree of mechanical interlocking that increases the friction between the seat and ball. This is particularly troublesome for fully Teflon lined ball and plug valves.

The amount of torque required by a ball valve also varies with the position of the ball. The maximum valve torque, called the breakaway torque, occurs at the start of the opening stroke, when the ball first starts to move. This is because the maximum pressure differential across the valve only occurs when the ball port is closed and no flow exists. Once the ball port becomes exposed to the fluid, flow starts and the differential pressure decreases. At this point, the torque level drops to its lowest point, known as running torque. The running torque of a ball valve is approximately one-third of the breakaway torque.

When gathering torque sizing information for any quarter-turn valve, it is important to know the following valve information.

1. Type of valve (ball, plug, butterfly)
2. Maximum torque and positions at which it occurs
3. Media flowing through the valve and maximum expected line pressure

In general, quarter-turn valves have what may be considered characteristic torque curves. The three major valve types—ball, butterfly, and plug—are covered in the following examples.

Characteristic torque curves for rotary valves

Ball valves. The most widely used ball valve design in the marketplace uses a polished sphere with a through port (ball) and a matching seat that seals off the flow. Shutoff occurs as the ball presses up against the seat due to the upstream pressure of the fluid. From this pressure-assisted sealing design it is clear that as the pressure increases, the amount of torque required to rotate the ball also increases. A ball valve has a torque curve that is characteristic of its design.

Some ball valve manufacturers size actuators based on two different curves—one for 1½ in (40 mm) valves and larger and one for smaller valves.

Curve 1 (Fig. 6.5), for 1½ in (40 mm) and larger valves, indicates that some manufacturers consider that the closing torque (operating from open to close) is 80 percent of the unseating torque (usually the published torque). Curve 2, for valves smaller than 1½ in (40 mm), indicates that both values are considered equal.

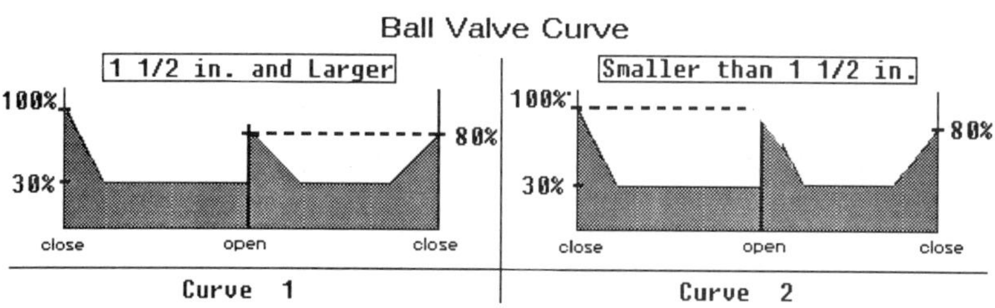

Figure 6.5 Typical operating torque curve for small- and large-diameter ball valves.

Butterfly valves. A butterfly valve closes off the flow of a fluid by squeezing a rotating disc into a rubber seat. This closed position is the highest point on the torque curve, as it requires the most turning force (see Fig. 6.6). In the open (full flow) condition, the disc is positioned parallel to the flow. Here the torque is low. After only a few degrees of rotation in the opening direction, flow begins around the disc, reducing the differential pressure and consequently the torque required at the stem.

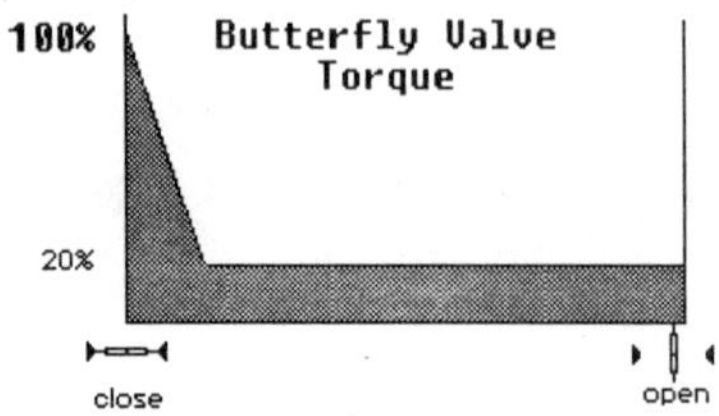

Figure 6.6 Typical operating torque curve for butterfly valve.

Torque requirements for a butterfly valve are determined by static forces, caused by the differential pressure across the disc in the closed position, and by dynamic forces, caused by fluid velocity in the pipe and at the edge of the partly closed disc as well as the inherent bearing-shaft-seat friction.

Some manufacturers of rubber-seated butterfly valves offer an undercut disc option. The disc diameter is reduced, which in turn reduces both the shutoff pressure rating of the valve and the seating torque. Smaller, more economical automation packages are possible when using this option.

Plug valves. Plug valves are designed with a tapered plug, which serves as the closure member, surrounded by a sleeve of sealing material, usually Teflon, to provide shutoff. This is known as a nonlubricated plug valve. It is not a pressure-assisted valve. Instead, the plug is continuously forced downward into the body against the seat, the taper of the plug acting like a wedge to provide sealing.

The continuous pressure on the plug and its taper design means that high torque is required throughout rotation to cycle the valve (see Fig. 6.7). Once the plug is in motion, the torque required to continue turning it is reduced by about 10 percent. For all valves, the portion of a torque curve between open and closed is called the *running* torque.

Dynamic torque. Dynamic torque is the rotary force required to oppose any fluid forces within the valve under flowing conditions. As fluid

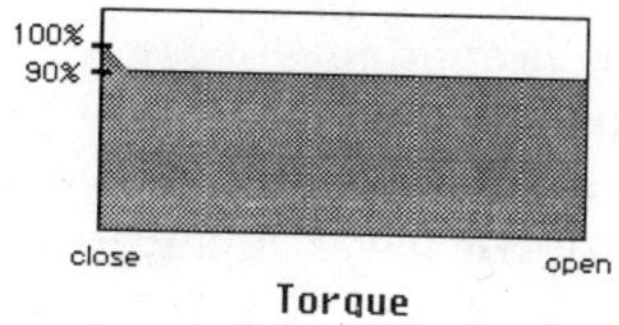

Figure 6.7 Typical operating torque curve for plug valve.

flows around the ball or disc, this pressure ultimately creates areas of high and low pressure on the ball or disc. This pressure results in increased torque on the valve shaft. Dynamic torque may tend to either close or open the valve depending on the direction of flow and the type of valve used.

If the tendency of fluid flow to either open or close the valve is extreme, positioning accuracy for a rotary-control valve may be diminished. When dynamic torque is a concern, the actuator output torque curve and valve curve should be properly sized for best operation.

6.3 Pneumatic Actuator Selection for Linear-Shaft Valves

Though rotary-shaft valves are widely preferred throughout the process industries because of the relative ease by which they can be automated, there are still some applications where automated gate valves or other linear-shaft valves are better suited. Selecting pneumatic actuators and accessories for linear-shaft valves is similar to the process used for rotary-shaft valves except that sizing is slightly different. In addition, where readily adaptable accessories for rotary pneumatic actuators are widely available, most linear actuator accessories must be fabricated to the assembly. Finally, the overall package is more predictable, for no matter what size the actuator is, all accessories are designed for 90° of rotation. Linear actuators, on the other hand, are selected based on stroke length which varies for each valve size. This makes accessory design a new challenge for each new assembly.

Two types of pneumatic actuators are used with linear valves to produce the necessary operating thrust for efficient operation, the piston and the diaphragm. Both generate output force (thrust) based on the formula

$$\text{Force} = \text{pressure} \times \text{area}$$

Once the necessary operating force for the valve is determined, the actuator size becomes a function of the supply pressure available. If for instance a supply of 100 lb/in^2 gage (7 bar) is available to generate 5000 lb (22 kN) of thrust, then the effective area of the piston must be at least 50 in^2 (325 cm^2). From the formula Area = πr^2 the resulting piston diameter would be about 8 in (200 mm). For lower supply pressures a piston with a larger diameter is necessary. At 50 lb/in^2 gage (3.5 bar) supply, for instance, a 16-in (400-mm) piston is required. This is often where diaphragm actuators are preferred.

To minimize the actuator size higher pressures may be used. For these applications high hydraulic pressure is recommended. These systems produce as much as 3000 to 4000 lb/in^2 gage (200 to 275 bar). Where high thrust valves require 50,000 to 100,000 lb (200 to 400 kN) or more, hydraulic piston actuators can provide reliable service in a compact design.

For example, if valve thrust = 100,000 lb (450 kN) and supply pressure = 4000 lb/in^2 gage hydraulic (27.6 MPa), then

$$\text{Area} = \frac{100{,}000}{4000} = 25 \text{ in}^2 \text{ (162 cm}^2\text{)}$$

$$\text{Piston diameter} = 5.6 \text{ in (6 in nominal) (150 mm)}$$

Linear piston actuators are widely available from many different fluid power companies. Adapting them to a linear valve is often more complicated than it may seem initially and should be performed by a qualified automation specialist or by the valve manufacturer.

Spring-and-diaphragm actuators for linear valves

Where piston actuators for linear valves may be double-acting or spring-return, diaphragm actuators are often used as a spring-and-diaphragm style for control applications.

From the force formula noted earlier it can be seen that very large actuator force requirements can be met by increasing diaphragm area, increasing supply pressure, or both. Obviously, use of larger diaphragms will require larger and heavier actuators, and there is a practical end limit to the size and weight of an actuator that can be used. Some of the largest spring-and-diaphragm actuators available use a diaphragm of between 400 and 500 in^2 (2600 to 2300 cm^2), are up to 10 ft (3 m) tall, and weigh in the neighborhood of 1500 lb (680 kg). Supply pressure for a diaphragm actuator is usually limited to about 65 lb/in^2 gage (4.5 bar). Maximum force available from the larger spring-and-diaphragm actuators is generally in the range of 10,000 lb (45 kN). Large spring-and-diaphragm actuators may require additional pipeline supports or special bracketry for mounting.

Actuator sizing for linear valves

With few exceptions gate and globe valve manufacturers do not publish thrust figures for actuator sizing. Consequently, very complex and conservative factors in thrust calculations may add significantly to an actuator's cost. Most valve manufacturers, however, will generally provide a recommended sizing value based on the application parameters. When these data are not available, there are two methods that may be used for actuator sizing. The first is a conservative estimate, and the second is more precise, based on having specific data on all the critical variables.

For a conservative estimate the following information is necessary.

1. Valve type
2. Valve size
3. Maximum differential pressure (the inlet pressure less the outlet pressure), which usually occurs in the closed position for on-off valves
4. Minimum pneumatic or hydraulic pressure available at the actuator
5. Valve stem diameter (Each actuator has a maximum allowable stem diameter.)

This information is necessary to calculate the amount of thrust that a linear actuator must produce to achieve the opening and closing of the valve.

The estimate formula is

$$\text{Gate valve thrust} = (\text{valve size})^2 \times \text{differential pressure} \times \text{valve factor} \quad (\text{VF} = 0.3)$$

$$\text{Globe valve thrust} = \text{port area} \times \text{differential pressure} \times \text{valve factor} \quad (\text{VF} = 1.25)$$

The precise method should always be used when possible. Given the variables, the proper thrust and actuator size may be calculated using the following procedure. First, the precise method requires knowledge of the valve manufacturer, model, type, and size and the valve port and stem diameters. Second, the following questions should be asked:

1. Is the valve a cage type, or does it have a double-guided stem?
2. Is the valve being used for on-off or throttling service?
3. If it is a gate valve, does it have a solid, split, flex, or parallel disc?

4. Is the seat metal-to-metal or soft? If a soft seat, what is the material?
5. Is the medium a liquid or a gas, and what are the temperatures?
6. What are the maximum inlet and minimum outlet pressures?
7. In what position is the valve to fail (as-is, open, or closed) in the event of air or electric failure?
8. Must the actuator stroke the valve in a specific period of time?
9. Are there any installed space limitations?
10. What controls and control interfaces are required?

The precise formula is

$$\text{Valve thrust} = \text{port area} \times \text{differential pressure} \times \text{VF} + \text{SB} + \text{SaT}$$

Valve factors

Gates	
Split and flex wedge	0.25
Solid wedge	0.30
Parallel disc	0.20
Teflon-seated	0.18
Liquids over 750°F (400°C)	0.30
Gases under 1000°F (540°C)	0.35
Gases over 1000°F (540°C)	0.40
Globe and angle valves	1.15

SB factors (stuffing box friction)

Stem diameter, in (mm)	Friction, lb (kN)
Up to 1 (25)	1000 (4.5)
1⅛–1½ (28–40)	1500 (6.8)
1⅝–2½ (40–65)	2500 (11.4)
2⅝–4 (65–100)	4000 (18)
4⅛–6 (100–150)	5000 (23)

SaT factors (pressure effect on stem)

$$\text{SaT} = \text{stem area} \times \text{maximum differential pressure}$$

Actuator thrust outputs

Valves in which the closure element moves to and from the seat, such as globe, angle, diaphragm, and wedge gate valves, depend to some degree on the mechanical force of the stem holding the closure against the seat to make and maintain tight shutoff. This is most important if the pressure to be shut off acts on the closure element in a direction to push it off the seat. In globe-type valves, when pressure acts in the same direction as the stem force, as well as in gate valves, pressure acts to assist seating load. In such valves stem loading is less critical, except that the stem force at low pressures may be more important than at high pressures.

Double-acting piston. The thrust output of a double-acting piston actuator is constant throughout the stroke in one direction. The output force (thrust) of a pneumatic piston is determined by using the following formula.

$$\text{Force} = \text{pressure} \times \text{area}$$

where force = thrust
pressure = pneumatic or hydraulic pressure applied to piston
area = area of piston = πr^2 (r = radius)

In the closed-to-open stroke when the piston rod is retracting, the effective piston area (the area upon which the supply pressure may act) is reduced by the area of the piston rod; therefore, the thrust output in the closed-to-open (retracting) stroke is less than the thrust output in the open-to-closed (extending) stroke. For typical low air pressures of 40 to 100 lb/in^2 gage (3 to 7 bar) the higher closing thrust is an advantage in seating the valve; however, high actuating pressures of 500 to 3000 lb/in^2 gage (35 to 200 bar) provide significantly higher closing thrusts versus opening thrusts. For high pressures it is necessary to use mechanical stops on the actuator to avoid driving and jamming the valve disc or plug into the seat.

Calculating cylinder thrust output on retracting stroke. Given: Piston diameter = 10 in (250 mm) and piston rod diameter = 2 in (50 mm).

Calculated: Area of piston = πr^2 = 78.5 in^2 (500 cm^2), area of piston rod = 3.14 in^2 (20 cm^2), and effective area = 75.36 in^2 (480 cm^2).

Spring-return piston. The thrust output of a spring-return actuator decreases during the air stroke, since the force is being expended to compress the spring, or on the return stroke, since the spring compression is being released. The beginning (break) thrust output is approximately 2 times greater than the ending thrust output. Also, since the piston area is larger for any given actuator size when failing to open, depend on the actuating pressure, a fail-open actuator may be smaller than a fail-closed actuator. In sizing for a valve that is to fail closed, make certain that there is sufficient thrust at the spring end, and for a valve that is to fail open, make certain that there is sufficient thrust at the pneumatic or hydraulic end.

The thrust outputs from spring-return actuators are ideally 67 percent of a double acting at the spring break and 33 percent at the spring end. This ratio provides perfect balance of both the spring and the pressure thrust outputs. This ideal is dependent on several factors.

1. Actuator operating pressure
2. Valve stroke
3. Actuator piston area
4. Available spring diameters
5. Available spring wire diameters
6. Available spring lengths

It is unrealistic and uneconomical for actuator manufacturers to offer an ideal spring for each set of conditions, so some elect to offer a thrust range for each model series from which sizing approximations can be made.

Sizing spring-return piston actuators for linear valves. To obtain the approximate spring-return actuator size, use 65 percent of the double-acting thrust for both the spring break and the pressure break, and use 30 percent of the double-acting thrust for both the spring end and the pressure end. The actual values could vary by ±10 percent from these approximate thrust outputs.

The fail-safe alternative. An economical alternative to a spring-return actuator is a fail-safe system, which uses a double-acting actuator and an air receiver. The receiver (or accumulator) stores high-pressure air and automatically releases it to the actuator upon loss of supply air, to fail the valve in either the open or closed position. The supply air flows through a check valve, to the receiver, to a four-way valve, and then to the actuator. The receiver is under a constant charge by the supply air. The controls can be configured so as to fail safe on loss of air, electric power, or both. Under normal conditions the actuator operates as a double-acting unit and provides maximum, constant thrust throughout the stroke in either direction. In the failure mode the receiver starts at full pressure, but since the volume of the system increases as the actuator cycles, the pressure, and consequently the available thrust, decreases as the actuator strokes. The amount of pressure reduction is related directly to the size of the receiver and the displaced volume of the actuator. The larger the receiver versus the actuator, the higher the end pressure and end thrust.

In sizing an accumulator the following data are required.

1. The valve thrust required at the beginning and end of the stroke
2. Whether the valve is to fail open or closed
3. The air supply to the actuator
4. The volume or displacement of the actuator [area × (valve stroke + 2 in)]
5. The length of valve stroke
6. If more than one, the number of strokes desired

Sizing procedure. The following example will provide a simple method for understanding the procedure. You are given a 10-in (250-mm) 600# gate valve, with a stroke of 11 in (280 mm), a stem diameter of 2 in (50 mm), and a required thrust of 15,250 lb (67,800 N) due to a differential pressure of 500 lb/in^2 (35 bar). The valve is to fail closed, and the supply air is 80 lb/in^2 (6 bar).

Step 1. With a 20 percent safety factor the sizing thrust will be 18,300 lb (81,400 N).

Step 2. With an 80 lb/in^2 (6 bar) supply pressure the piston diameter must be at least 17.07 in (430 mm) [an 18-in (460-mm) diameter piston has 254 in^2 (1650 cm^2)].

Step 3. Since we require at least 15,250 lb (67,800 N) of thrust for full valve closure, we can divide thrust by area and get the minimum air-supply pressure: 15,250/254 = 60 lb/in^2 (4 bar).

Step 4. From Fig. 6.8 we can select the air receiver factor by going down the 60 column to the 80 row. The tank must be 4.79 times the actuator volume. The actuator volume is 254 in^2 × 11 in = 2794 in^3 (45 L). The tank must be 2794 × 4.79 = 13,383 in^3 (220 L).

Accessories for linear pneumatic actuators. Some accessories typically available for linear pneumatic actuators include

Adjustable opening stop. The external threaded bolt can be adjusted to limit the opening of the valve. In an emergency, smaller valves can be closed using this device.

Adjustable closure stop. An adjustable closing travel stop will prevent overclosure of the valve due to the use of excessive air pressure.

Position indicator. A metal rod enclosed in a plastic tube indicates whether the valve is open or closed. Position indicators can be fur-

Table 1* Air receiver volume (x) vs actuator volume (1.0)

P1 \ P2	95	90	85	80	75	70	65	60	55	50	45	40	35	30	25	20
100	27.54	13.21	8.42	6.03	4.60	3.65	2.96	2.45	2.05	1.73	1.46	1.24	1.05	.89	.75	.63
90			25.05	11.96	7.60	5.41	4.10	3.23	2.60	2.13	1.77	1.47	1.23	1.03	.86	.71
80					22.55	10.71	6.77	4.79	3.60	2.81	2.24	1.82	1.49	1.22	1.00	.82
70							20.05	9.46	5.93	4.16	3.10	2.39	1.88	1.50	1.20	.96
60									17.55	8.21	5.10	3.54	2.60	1.97	1.52	1.18
50											15.05	6.96	4.26	2.91	2.10	1.55
40													12.55	5.71	3.43	2.28

*Air receiver volume must be X times larger than actuator volume for 1 stroke of actuator.

Figure 6.8 Air receiver volume (x) versus actuator volume (1.0). (*Courtesy of Automax Inc.*)

nished for field installation if necessary. For larger valves an arrow indicator may be attached to the valve stem or actuator rod.

Limit switches. Available in all NEMA standards. In most cases limit switches can be field-installed. Proximity switches are also available.

Positioner. For automatic throttling control. Box-type positioners for side mounting require a yoke-type bonnet.

Accumulator tank. To provide for an air reserve for driving a double-acting actuator to a desired fail position.

6.4 Pneumatic Actuators and Controls for Rotary-Shaft Valves

There have been a number of studies done and articles written about the key position valve actuators will hold in the process systems of the future. Much of the growth of rotary actuator use is related directly to the increased use of rotary (quarter-turn) valves.

Rotary valves (ball, plug, and butterfly valves) continue to be the fastest-growing segment of the entire control valve industry. This is because they

1. Are inherently reliable
2. Are lightweight
3. Are low in cost
4. Are easy to maintain
5. Feature high flow capacity
6. Are easy to automate
7. Come in a wide variety of materials

Once thought to be a valve accessory, actuators are now considered an integral part of the flow control function.

Selecting the right actuator is as important as picking the correct valve. An actuator evaluation should include a review of product literature and, if possible, a visual inspection of the actuator. Understanding the characteristic torque output curves will help ensure the best valve performance.

The actuator size and weight are critical if the valve must meet seismic specifications. Actuators should be supplied with valve position indicators. They should be readily visible and not easily damaged.

It has generally been the responsibility of the valve manufacturer to select and specify the valve and actuator for a given application because the manufacturer has the most complete and detailed knowledge of valve performance and the particulars of the valve for actuator sizing. Insufficient understanding of the service and of how to relay the necessary information to the valve manufacturer has penalized valve users in some cases.

Today, there is an abundance of information available to help the user size and select actuators based on a wide variety of conditions. This means that the user may now specify a valve that best meets the process requirements (flow rate, materials, pressure and temperature rating) and then select an actuator with control accessories that match the service (high cycle, two-stage shutoff) and control system. Before the selection and sizing of an actuator may begin, however, valve performance and torque under very specific operating conditions must be determined.

Pneumatic actuators for rotary valves in on-off service may be classified by the supply air used by each for operation—low pressure and high pressure. Low-pressure actuators operate in a supply pressure range of about 20 to 45 lb/in^2 (1 to 3 bar), and include diaphragm, spring and diaphragm, and some piston actuators. Since rotary spring and diaphragm actuators operate on relatively low pressure, diaphragm sizes must be quite large to obtain usable outputs. Actuator size and weight may be a problem at installation and may require floor stands or other support to prevent putting too much stress on the valve neck and piping system. High-pressure actuators operate in a supply pressure range of about 40 to 150 lb/in^2 gage (10 bar), and include piston actuators, vane actuators, and some diaphragm actuators. The rotary pneumatic actuator, due to its inherent torque characteristics, can greatly simplify valve and actuator sizing and provide power for throttling as well as on-off valve applications.

The following factors are important to consider when selecting a rotary pneumatic valve actuator.

1. Actuator compatibility with the proposed or existing control systems
2. Torque characteristics of the actuator and torque requirements of the valve
3. Control accessories and options that will adapt to and work with the actuator and the process control instrumentation
4. Emergency shutdown mode (methods for fail-safe requirements)
5. Operating environment (hazardous, corrosive, cold, hot, humid)
6. Physical size of the actuator and installation methods

A rotary pneumatic actuator may be defined as an air-powered device that produces torque (turning force) to rotate the stem (and thus the closure member) of a valve for flow control. The output of a pneumatic rotary actuator is torque. Torque is defined by the formula:

$$\text{Torque} = \text{force} \times \text{ma}$$

where force = piston force, defined as $p \times A$
p = air pressure
A = piston area
ma = moment arm. This is the distance from center of rotation to where force is applied

Now the torque formula may be extended to read

$$\text{Torque} = \text{pressure} \times \text{area} \times \text{moment arm}$$

Since force is given in pounds (newtons) and moment arm in inches (meters), the product, torque, is given in pound-inches (newton-meters). This may be converted to pound-feet by dividing by 12.

From this formula it is clear that to increase the amount of torque produced by a rotary actuator it is necessary to increase the air pressure p to the cylinder, increase the piston diameter which increases the piston area A, and/or increase the moment arm dimension ma. For this reason actuator manufacturers will produce actuators in a variety of models with physical dimensions that increase the piston area, moment arm, and overall output torque range. Since each model will produce output torque based on a range of supply pressure, a typical torque chart will show increasing torque values as supply air pressure increases.

Early pneumatic actuator mechanisms

Some of the early actuators first introduced to the industry were simply products of a hasty response to the process industry's requirements for more valve control options. When the need for automatically controlled rotary valves expanded, manufacturers, who wished to participate in the growth, developed actuators by collecting and assembling readily available pneumatic, mechanical, and control components.

Linear air cylinders, or diaphragms, were adapted to the rotary stem of a valve with clevis pins and fabricated linkage assemblies (see Fig. 6.9). Control accessories, such as solenoid valves and limit switches, were adapted to the actuators with plates, fasteners, or welded components. These makeshift assemblies were inefficient, and the piston rods and linkage assemblies of these actuators were exposed to the environment, causing corrosion damage, seal failure, and

Figure 6.9 Early actuator designs included the linear air cylinder.

general unreliability. Finally, the linkage design was a potential hazard to plant personnel. When the valve was actuated, a scissoring action was created by the linkage as the cylinder would pivot, making work near the assembly dangerous.

In overcoming the problems associated with these early designs, actuator manufacturers and some valve manufacturers have developed mechanisms that are more suitable for plantwide valve automation.

Current actuator designs

Rotary pneumatic actuators consist of basically two elements: a force-generating device (diaphragm or piston) and a linear-to-rotary conversion mechanism. Selecting the best design for an application will ensure dependable performance.

Diaphragm actuators. The diaphragm is a piece of fabric-reinforced elastomer that is sandwiched between two halves of a casing. Connected to the center of the diaphragm is a rod that is part of the linear-to-rotary linkage. Pressurizing one side of the diaphragm generates a force F on the rod which is transmitted to a crank arm or lever which serves as the moment arm ma. The resultant rotary force is, of course,

Figure 6.10 Cutaway view of spring-diaphragm actuator. (*Courtesy of Neles-Jamesbury, Inc.*)

torque. When the diaphragm is depressurized, spring force acts to return the valve to its original position (see Fig. 6.10).

There are three basic diaphragm shapes: flat, molded, and rolling. The flat diaphragm's stroke length is limited, and its effective area changes with stroke position. This reduces the actuator's torque capacity and restricts it to use on small valves that are not in throttling service. The molded diaphragm is dished and provides for longer travel and a more-uniform effective area. The rolling diaphragm permits even longer strokes and essentially eliminates changes in the effective area; however, it can only be pressured in one direction. Both the molded and rolling types are suitable for throttling and on-off service.

One feature of the spring and diaphragm actuator is its ability to provide relatively linear output response with respect to an increasing supply pressure. For example, in an actuator operated with a 3- to 15-lb/in^2 gage (0.2- to 1-bar) input signal, the midpoint of actuator stem travel would occur at 9 lb/in^2 gage (0.6 bar) input. The valve closure element would then be approximately halfway between the fully open and fully closed positions. In many applications, diaphragm actuators with linear response can eliminate the need for a positioner.

Piston actuators. The piston actuator consists of a cylinder and a piston that is normally sealed with an elastomer O-ring. Piston actuators may be used as a double-acting (air pressure to one or the other side of the piston to cycle a valve) or spring-return device.

They may be operated at air-supply pressures more than twice the maximum allowed by spring-diaphragm actuators making them more powerful and compact. In control applications piston actuators must be installed with a positioner.

Torque-transfer mechanisms. Because piston and diaphragm actuators generate linear force (thrust) they must be used with some type of linear-to-rotary conversion mechanism (torque-transfer) for use with rotary valves. An actuator's suitability for a service depends considerably on the mechanism by which the linearly applied force from the piston or diaphragm is transmitted to the valve shaft. The conversion from linear to rotary force may be accomplished in a number of ways. We shall cover three popular pneumatic actuator designs, detailing the mechanism and providing a basis for evaluation. They are the scotch-yoke, vane, and rack-and-pinion actuators.

Scotch-yoke actuators. The scotch-yoke piston actuator is used for rotary operation of quarter-turn valves. The mechanism consists of a diaphragm (or piston), power-transfer pin, yoke, and output shaft (see Fig. 6.11). When air pressure is applied to the piston, a linear force is generated ($p \times A$). This force is applied to the yoke, through the power-transfer pin, at a distance ma from the center of the output shaft. As the yoke moves in a rotary fashion, torque is generated at the output shaft to provide the turning force required to cycle a valve.

To measure the output torque of a scotch-yoke actuator it is necessary, holding p and A constant, to measure the distance ma from the point of applied power to the center of rotation at all points in the piston stroke. At both extremes of travel (open and closed), ma is greatest. At the midpoint of travel, ma is at its shortest dimension. Substituting these various dimensions into the torque formula produces the

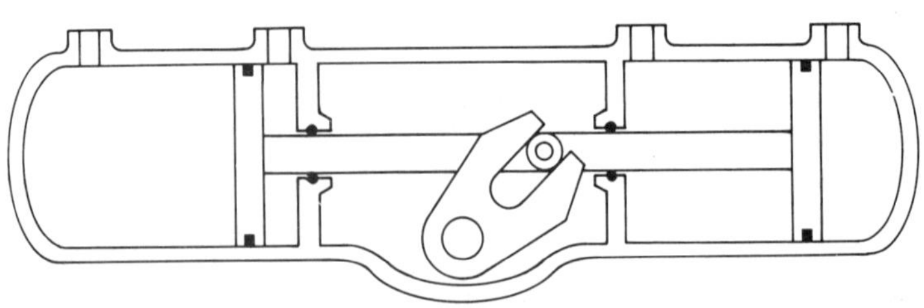

Figure 6.11 View of piston actuator with scotch-yoke mechanisms.

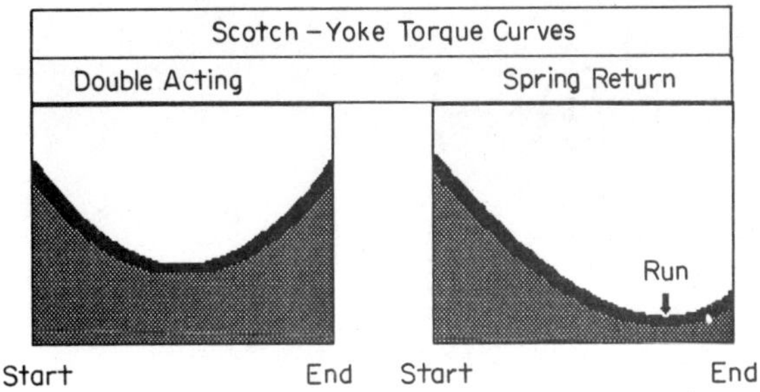

Figure 6.12 Output torque curve of scotch-yoke actuator.

following output torque curve for a scotch-yoke actuator over 90° of rotation (see Fig. 6.12).

This is the characteristic output torque curve of a double-acting scotch-yoke actuator as it strokes from open to closed (0 to 90°). A torque chart for a double-acting scotch-yoke actuator will often list two values for sizing: breakaway, the highest value; and running, the lowest. Spring-return torque charts usually give the start and end values.

Vane actuators. Vane actuators are simple quarter-turn devices. The design uses a vane (or paddle) that is fitted with an O-ring, a molded rubber coating, or other sealing system. The vane has an integral output shaft, which is used to cycle a valve (see Fig. 6.13). When air is applied to the actuator, it acts upon the area of the vane to produce force at the effective center of the vane. This force point may be calculated based on the dimensions of the vane, but it is the point, when measured from the center of rotary output, where ma is determined. As the vane cycles about the center of the output shaft, ma remains constant, so the output torque curve of a double-acting vane actuator for a full cycle is constant. During operation, the vane sweeps and slides the paddle seal along the inner wall of the housing. The stroking distance that the seal sees is equal to the length of the arc at the point of seal and housing contact and should be considered when high cycle life is a requirement.

Because there is no linear-to-rotary mechanism the task of creating a fail-safe vane actuator has been approached in many different ways. Three common methods are the coil spring module, the linear-to-rotary spring module, and the accumulator tank system (see Fig. 9.1). The first two are mounted to the top or bottom of the actuator housing.

Rack-and-pinion actuators. The pneumatic rack-and-pinion design has become widely accepted and is available for virtually any range of air pressure, torque, valve size, and type. The rack-and-pinion mecha-

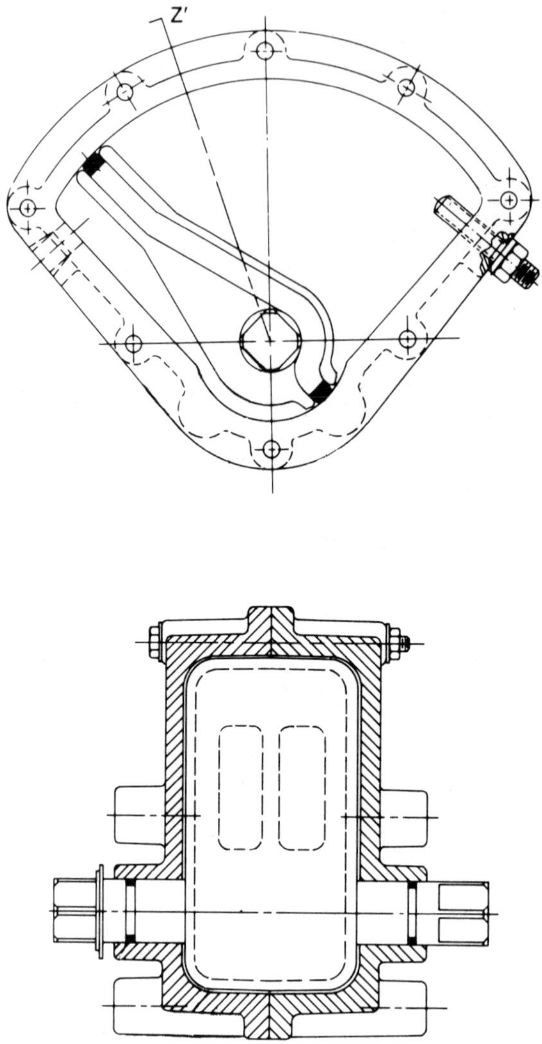

Figure 6.13 Vane actuator. (*Courtesy of Neles-Jamesbury, Inc.*)

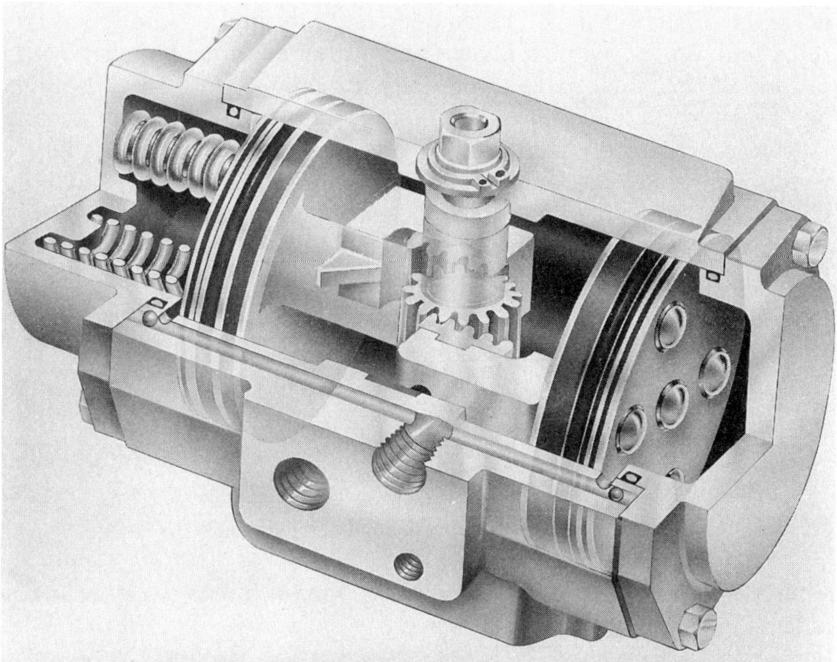

Figure 6.14 Cutaway of rack-and-pinion actuator shows spring-return and double-acting end caps. (*Courtesy of El-o-matic, Inc.*)

nism consists of a length of linear gearing (rack) and pinion gear, which serves as the rotary output shaft (see Fig. 6.14). The gear rack is fitted to or machined directly into an arm that extends from the face of the piston. As the piston strokes, the gear rack engages and applies force to the pinion at a dimension ma from the center of rotary output, which is the radius of the pinion shaft. This produces torque.

The stroking length of the piston is limited, and designed to produce output torque over 90° (one-quarter turn) of rotation. Rack-and-pinion actuator options include 180° rotation for multiported valves.

The gear rack contacts the pinion at a point that is a constant dimension from the center of rotation; therefore, the torque output for a double-acting actuator throughout rotation is constant.

The torque characteristic curve for a double-acting rack-and-pinion actuator is a straight line. For this reason, a torque table for double-acting rack-and-pinion actuators will show one value for each model size at a given operating pressure. This simplifies the process of actuator selection for valve and actuator sizing and also represents a torque advantage over other types of actuators.

A rack-and-pinion actuator may be manufactured as a single-piston or double-piston design. The double-piston design is usually referred to as a dual-opposed piston rack-and-pinion actuator because, as the pistons stroke, they move in opposite directions—inward and toward one another in one direction and apart in the other.

The dual-opposed piston design provides more torque than a single-piston design with the same cylinder diameter because air pressure is applied to two piston areas simultaneously, effectively doubling the area A in the torque formula without increasing the piston diameter. Three configurations of the rack-and-pinion design are shown in Fig. 6.15.

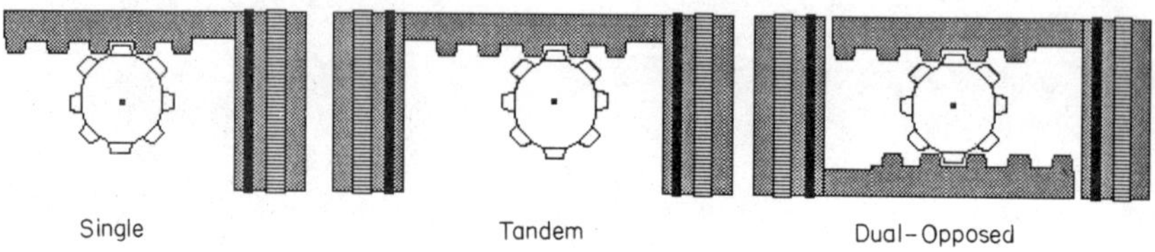

Figure 6.15 Three configurations of the rack-and-pinion design.

Other linear-to-rotary mechanisms. Other types of linear-to-rotary mechanisms for pneumatic actuators are

1. *Direct-coupling.* The piston rod is connected directly to the crank arm. The torque output curve is the inverse of the scotch-yoke. Maximum output occurs at the midposition (45°).
2. *Link-coupling.* A short link connects the piston rod to the crank arm. The rod travels linearly allowing for a simple rod-sealing arrangement.
3. *Helical-cam.* This vertically assembled actuator is often compared to the "yankee screwdriver." The piston rod is fitted with a pin that engages a slot in the output member to generate rotary force. Wear of the mechanism is similar to that of the scotch-yoke design. Output torque is constant throughout the stroke.

Actuator sizing techniques

There are two characteristics in the relationship of torque output versus rotational position in double-acting pneumatic actuators—variable and constant.

Variable output torque is produced when a *constant* linear piston force ($p \times A$) is applied to a *varying* moment arm (ma). Actuators using a scotch-yoke, link-coupling, or direct-coupling mechanism generate variable output torque over their rotary output range.

Rack-and-pinion actuators and vane actuators have a constant torque produced by applying *constant* linear force to a *constant* moment arm dimension. The result is a curve that has a high torque value at all points throughout rotation. If properly sized to exceed the maximum torque required at the valve stem, a double-acting rack-and-pinion actuator, because of its constant power output, will exceed the torque required at any point in travel. Likewise, in throttling applications where the intermediate positioning force may be increased by fluid velocity or extended set periods, these actuators provide constant torque, fast response, and good controllability.

Valve torque characteristics

Opening and closing a rotary valve requires torque. Sizing is the precise matching of an actuator with the torque requirements of a valve. If an actuator is too powerful (oversized) for a valve, it can damage the valve by forcing the stem when it or the closure member is obstructed. If undersized, the actuator will not produce enough torque to operate the valve. Not being able to close the valve is a serious problem, particularly with spring-return (fail-safe) actuators. Proper sizing of a valve actuator means longer, more dependable service, less maintenance, and economical operation.

When selecting an actuator for valve automation, it is important to consider that each of the most common types of quarter-turn valves (ball, plug, and butterfly valves) have unique torque curves that are characteristic of their design and that the torque required to rotate a valve from open to closed may fluctuate at almost any point in its travel, because of a number of factors including:

1. Line pressure
2. Stem-packing tightness
3. Set period
4. Temperature
5. Viscous or slurry media
6. Age
7. Fluid velocity

Some manufacturers will even suggest sizing safety factors when clean, clear fluids aren't used or for any other unusual flow or fluid conditions. Most manufacturers recommend the following sizing factors.

Conditions	Sizing factor, %
Dry gas or air	25 (50 for rubber-seated valves)
Slurry	25–50
High velocity	25
High pressure	25
High temperature	25
Infrequent cycling	25
Retrofitting	25

Thus, it is important to understand valve and actuator torque curves and actuator sizing methods when selecting an actuator for plantwide valve automation.

Most valve manufacturers will publish the maximum torque required under given conditions. Sizing a double-acting actuator with constant torque does not require that more details about the valve torque curve be given. It is important to consider that some valve types require greater operating torque than others for the same line size and service. For instance, 2-in (50-mm) ball, plug, and butterfly valves in 150# service require the following operating torques (values are intended as a guide only):

Valve	Operating torque, lb · in (N · m)
Ball	200 (23)
Plug	470 (53)
Butterfly (rubber-seated)	100 (11)
HP butterfly	190 (22)

From this you can see that for the same service some valves will require larger or smaller actuators. Of course, valves should be selected primarily on their unique ability to handle the service required. But if the valves are determined to be equally capable, the package size, price, and weight should be considered.

Sizing a double-acting actuator

To size a double-acting constant-torque actuator (i.e., rack-and-pinion or vane), select the appropriate actuator model size from the torque tables (Fig. 6.16) with a torque output at a given air pressure that exceeds the maximum published valve torque. Include a 10 percent sizing safety factor. In practice, air-supply pressure at the actuator ranges from 60 to 80 lb/in^2 gage (4 to 6 bar) and is generally less than compressor settings.

DOUBLE ACTING TORQUE

Actuator Size		Torque (in lbs.) vs Supply Pressure (psig)					
		20	40	60	80	100	120
1.5	DA	30	60	90	120	150	180
3.5	DA	80	160	240	320	400	480
6	DA	140	280	420	560	700	840
15	DA	360	720	1080	1440	1800	2160
28	DA	680	1360	2040	2720	3400	4080
50	DA	1200	2400	3600	4800	6000	7200
75	DA	1800	3600	5400	7200	9000	10800

Figure 6.16 Double-acting torque table for rack-and-pinion design. (*Courtesy of El-o-matic Inc.*)

Example Given a published valve torque (max.) of 925 lbf · in (105 N · m), we incorporate a 10 percent safety factor.

$$925 \text{ in} \cdot \text{lb } (100 \text{ N} \cdot \text{m}) \times 1.10 = 1018 \text{ lbf} \cdot \text{in } (115 \text{ N} \cdot \text{m})$$

Air-supply pressure is 60 lb/in^2 (4 bar).

answer Actuator = DA 15 [1080 lbf · in (122 N · m)].

The method for sizing a double-acting pneumatic actuator with variable output (scotch-yoke, linkage, etc.) is similar to sizing for spring return in that complete knowledge of the valve torque curve is important. For this reason a full understanding of spring-return actuator sizing should prove adequate for virtually all actuator sizing.

Spring-return actuators

Spring-return actuators are used in a process to cycle a valve to a predetermined (fail-safe) position (open or closed) upon loss of air pressure. In a spring-return actuator, air pressure powers the cylinder in one direction only. The actuator is returned to its original position by spring action. For this reason, it is sometimes referred to as a single-acting actuator.

Some pneumatic actuator manufacturers use a multiple-spring design on spring-return models. The effective spring rate and hence the resulting return force are determined by the number of springs installed against the face of each piston (see Fig. 6.17). Depending upon where an actuator is used in a plant or how many pneumatic controls or equipment are in operation, the available air pressure may range from 40 to 100 lb/in^2 (3 to 7 bar). The ability to vary the quantity of springs avoids having to use or stock single springs of different rates to accommodate air pressure variations.

Output torque on spring-return actuators is directly affected by the spring compression force, which is determined by the number of type

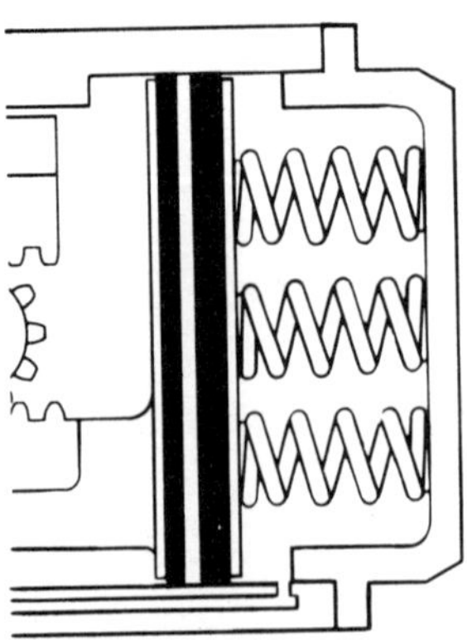

Figure 6.17 Multiple-spring actuator readily adapts to various supply air pressures. (*Courtesy of El-o-matic, Inc.*)

of springs used. Output torque decreases on the air stroke as springs are compressed and also on the spring stroke as they extend. This is because piston force is producing torque at the output shaft of the actuator but is also compressing the springs, which becomes increasingly difficult as they compress and thus reduces the effective output torque. Torque charts are used to determine precisely the correct actuator model and spring quantity for a given air pressure.

Spring-return sizing rules. When sizing spring-return actuators, there are four key values of output torque to consider. Two of these occur on the air stroke and two on the spring stroke. They are

1. Air start
2. Air end
3. Spring start
4. Spring end

(1 and 2: Air stroke; 3 and 4: Spring stroke)

They are located on the torque output graph shown in Fig. 6.18.

General rules for valve and actuator sizing for ball, butterfly, and plug valves are given here.

Ball valves. Ball valves generally require a higher breakaway (unseating) torque than running torque, but because frequently only one torque figure is published, the rated torque should be exceeded by the spring-end and air-start torques in spring-to-close applications and by the spring-start and air-end torques in spring-to-open applications.

Butterfly valves. Butterfly valves have a great variation in torque characteristics, depending upon the manufacturer and duty. Frequently, the breakaway torques (seating and unseating) are quoted, and these values should be exceeded by the spring-end and air-start

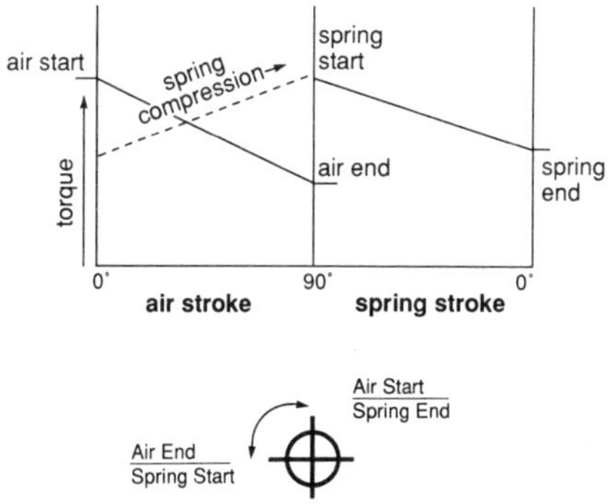

Figure 6.18 Spring-return torque curve for rack-and-pinion actuators.

torques for spring-to-close applications and by the spring-start and air-end torques for spring-to-open applications.

Plug valves. Plug valves are considered constant-torque valves because the running torque may be as much as 90 percent of the breakaway. The published torque value should be exceeded by the spring-end torque and air-end torque in all spring-return applications.

Spring-return sizing. As an example, let us size a spring-return actuator for each valve type considering two conditions: spring-to-close (fail closed) and spring-to-open (fail open). Refer to the torque tables (Fig. 6.19) for the following examples.

Ball valve—spring-to-close. Select the appropriate-size actuator from the torque tables with torque output at spring end and air start at a given air pressure exceeding the rated valve torque.

Example Given a published valve torque of 195 lb · in (22 N · m) and incorporating a 10 percent safety factor, torque required is

$$195 \text{ in} \cdot \text{lb } (22 \text{ N} \cdot \text{m}) \times 1.10 = 215 \text{ lbf} \cdot \text{in } (24 \text{ N} \cdot \text{m})$$

The given air-supply pressure is 80 lb/in^2 gage (6.0 bar).

1. Locate spring end: 223 lbf · in (25 N · m)
2. Locate air start: 313 lbf · in at 80 lb/in^2 gage (35 N · m at 6.0 bar)
3. Actuator model: SR-6-14

The last figure in this example, 14, is the quantity of springs used with the model SR-6, which is a multiple-spring design, to generate the amount of torque required.

Ball valve—spring-to-open. Select the appropriate-size actuator with torque output at spring start and air end at a given air pressure exceeding the rated valve torque.

Example Given a published valve torque of 195 lbf · in (22 N · m) and incorporating a 10 percent safety factor, torque required is

$$195 \text{ in} \cdot \text{lb } (22 \text{ N} \cdot \text{m}) \times 1.10 = 215 \text{ lbf} \cdot \text{in } (24 \text{ N} \cdot \text{m})$$

The given air-supply pressure is 80 lb/in^2 gage (6.0 bar).

1. Locate spring start: 230 lbf · in (26 N · m)
2. Locate air end: 306 lbf · in (35 N · m)
3. Actuator model: SR-6-8

Butterfly valve—spring-to-close. Select the appropriate-size actuator from the torque tables with torque output at spring end and air start at a given air pressure exceeding the rated valve torque.

SPRING RETURN TORQUE

Actuator Size	Number of Springs	40 PSI		60 PSI		80 PSI		100 PSI		120 PSI		Spring Torque	
		Air Start	Air End	Air Start	Air End	Air Start	Air End	Air Start	Air End	Air Start	Air End	Max Spring Start	Min Spring End
SR1,5	2			56.5	7.5	91.4	42.4	126.4	77.4	161.3	112.3	97	48
SR3.5	8	108.2	62.2	187.6	141.6	267	221	346	300	425.7	379.7	96	50
	10	96.2	38.2	175.6	117.6	255	197	334	276	413.7	355.7	120	62
	12	84.2	15.2	163.6	94.6	243	174	322	253	401.7	332.7	143	74
	14			150.6	70.6	230	150	309	229	388.7	308.7	167	87
SR6	8	140.7	37.7	275	172	409	306	543	440	678	575	230	127
	10			243	114	377	248	511	382	646	517	288	159
	12			211	57	345	191	479	325	614	460	345	191
	14					313	133	447	267	582	402	403	223
SR15	8	369.3	148.3	710	489	1051	830	1392	1171	1733	1512	531	310
	10	289.3	15.3	630	356	971	697	1312	1038	1653	1379	664	390
	12			550	223	891	564	1232	905	1573	1246	797	470
	14			480	91	821	432	1162	773	1503	1114	929	540
SR28	8	659	49	1301	691	1943	1333	2585	1975	3226	2616	1230	620
	10			1142	390	1784	1032	2426	1674	3067	2315	1531	779
	12			982	81	1625	723	2267	1365	2908	2006	1840	938
	14					1474	413	2116	1055	2757	1696	2150	1089
SR50	8			2479	1080	3615	2216	4751	3352	5887	4488	2319	920
	10			2240	496	3376	1632	4512	2768	5648	3904	2903	1159
	12					3146	1057	4282	2193	5418	3329	3478	1389
	14					2915	472	4051	1608	5187	2744	4063	1620

Figure 6.19 Spring-return torque table for rack-and-pinion actuator. (*Courtesy of El-o-matic Inc.*)

Example Given a published valve torque of 400 lbf · in (45 N · m) and incorporating a 10 percent safety factor, torque required is

$$400 \text{ in} \cdot \text{lb } (45 \text{ N} \cdot \text{m}) \times 1.10 = 440 \text{ lbf} \cdot \text{in } (50 \text{ N} \cdot \text{m})$$

The given air-supply pressure is 60 lb/in^2 gage (4 bar).

1. Locate spring end: 470 lbf · in (53 N · m)
2. Locate air start: 550 lbf · in (62 N · m)
3. Actuator model: SR-15-12

Butterfly valve—spring-to-open. Select the appropriate-size actuator from the torque tables with torque output at spring start and air end at a given air pressure exceeding the rated valve torque.

Example Given a published valve torque of 400 lbf · in (45 N · m) and incorporating a 10 percent safety factor, torque required is

$$400 \text{ in} \cdot \text{lb } (45 \text{ N} \cdot \text{m}) \times 1.10 = 440 \text{ lbf} \cdot \text{in } (50 \text{ N} \cdot \text{m})$$

The given air-supply pressure is 60 lb/in^2 gage (4 bar).

1. Locate spring start: 531 lbf · in (60 N · m)
2. Locate air end: 489 lbf · in (55 N · m)
3. Actuator model: SR-15-8

Plug valve. To size a spring-return actuator for a plug valve for either spring-to-close or spring-to-open, select the appropriate-size actuator from the torque table with a minimum torque output at any point in the cycle exceeding the rated torque of the valve.

Example Given a published valve torque of 675 lbf · in (76 N · m) and incorporating a 10 percent safety factor, torque required is

$$675 \text{ in} \cdot \text{lb } (76 \text{ N} \cdot \text{m}) \times 1.10 = 743 \text{ lbf} \cdot \text{in } (84 \text{ N} \cdot \text{m})$$

The given air-supply pressure is 80 lb/in^2 gage (6 bar). From this information we can locate the model: SR-28-10 [minimum output = 779 lbf · in (88 N · m)].

This is the smallest model size whose air stroke values [at 80 lb/in^2 gage (6 bar)] and spring stroke values *all* exceed 743 lbf · in (84 N · m).

Summary. The best method for sizing an actuator includes having all the particulars about the valve torque requirements at all points in the cycle. If this is not known, then assuming a constant-torque requirement throughout rotation for the valve is the next most appropriate and conservative step.

6.5 Control Accessories and Options for Pneumatic Actuators

Today's process controls range from complete computer systems to the staff-monitored electromechanical type (push buttons, heavy-duty relays, etc.). In the process area, there may also be pressure switches, temperature controllers, or other process-monitoring devices that must tie into the control valve and therefore the actuator.

When selecting a pneumatic rotary actuator for valve control in process applications, it is important that it be compatible with other components of the control system (power medium, control signals, etc.), the environment (corrosion, temperature), the system (speed, cycle frequency, fail mode), and, of course, the valve.

To work well with an existing control network, the actuator must be available with a few basic control accessories.

Solenoid valve. As a pilot device, available in various voltages and construction for the area classification

Limit switches. For indicating valve position, sequence cycling, alarms, etc.

Positioner. To throttle the valve in response to a varying control signal

Manual override. To operate the valve upon loss of air pressure or to simulate operation during startup

For environmental compatibility, the actuator should be available with corrosion-resistant (stainless-steel) trim, various coatings (polyurethane, epoxy, etc.), and weatherproof, hazardous-area, or intrinsically safe electrical accessories.

Pilot valve

A pilot valve for a pneumatic actuator is a control device that receives a manual or power signal and then directs air pressure to the air inlet ports of the actuator to drive it to the desired position. The most common type of pilot device is the solenoid-operated valve. As an electric device, it readily interfaces with widely used electromechanical control systems and may also be supplied with low-wattage coils for compatibility with computer control signals.

Pilot valves for pneumatic actuators are categorized by the number of port openings or ways air may flow. For instance a three-port (three-way) valve has a pressure port, output port, and exhaust port (see Fig. 6.20). The three-way valve is a logical choice for spring-return pneumatic actuators because only one air chamber is alternately pressurized or exhausted in normal operation. A four-way valve has a pressure port, two output ports, and an exhaust function (see Fig. 6.21). The two output ports will pressurize one or the other chambers of a double-acting cylinder, and so it is used with these types of pneumatic actuators.

These valves may also be described by the number of positions pro-

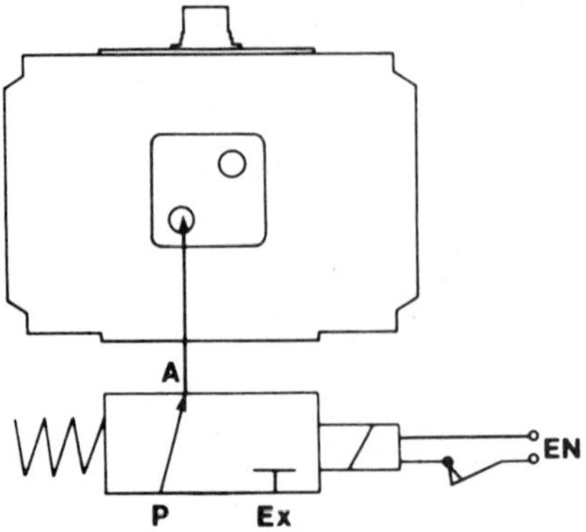

Figure 6.20 Pilot valve schematic (three-way) for spring-return actuator. (*Courtesy of El-o-matic, Inc.*)

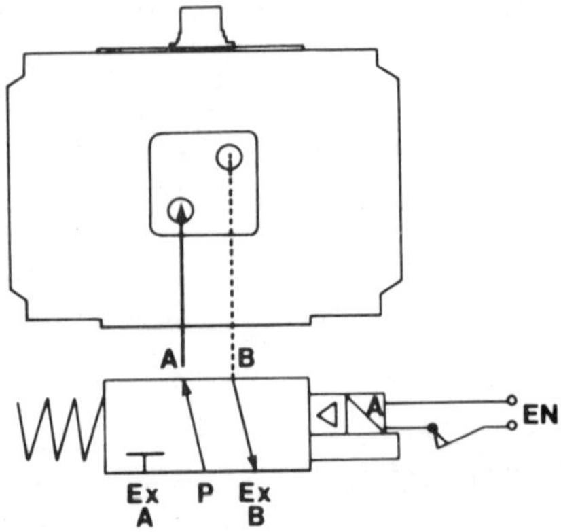

Figure 6.21 Pilot valve schematic (four-way) for double-acting actuator. (*Courtesy of El-o-matic, Inc.*)

vided. Typically their description gives the number of ports and number of positions as figures separated by a stroke—e.g., a 4/2 valve is a four-port, two-position valve. A five-way, two-position valve (5/2) is similar to a four-way valve except that an additional exhaust port is provided. Thus when either of the output ports is switched to exhaust, it does so through a separate exhaust port. This is of particular advantage when the stroke speed of a double-acting actuator must be controlled in both directions. Independent flow control valves may be connected to each exhaust port.

Another useful valve option is the four-way, three-position valve. The function is similar in porting to a four-way, two-position valve except that an additional position is available with all ports blocked. No flow is possible through the valve in either direction in the normal position. A typical application is the control of a double-acting pneumatic actuator that must be held in a given position upon loss of electric or air power. Another version of the 4/3 valve is to have the normal position provide air pressure to both sides of the actuator and the exhaust closed. This holds the actuator in position by pressurizing both sides of the cylinder.

Standard valve symbols

Symbols used for designating pilot valves are conventionally in the form of adjacent squares with each square representing one position of the valve. This means that a two-position valve would have two squares and a three-position valve would have three. Flow paths within the valve are indicated by lines and arrows (see Fig. 6.22).

The action of a pilot valve in an electric control circuit is initiated

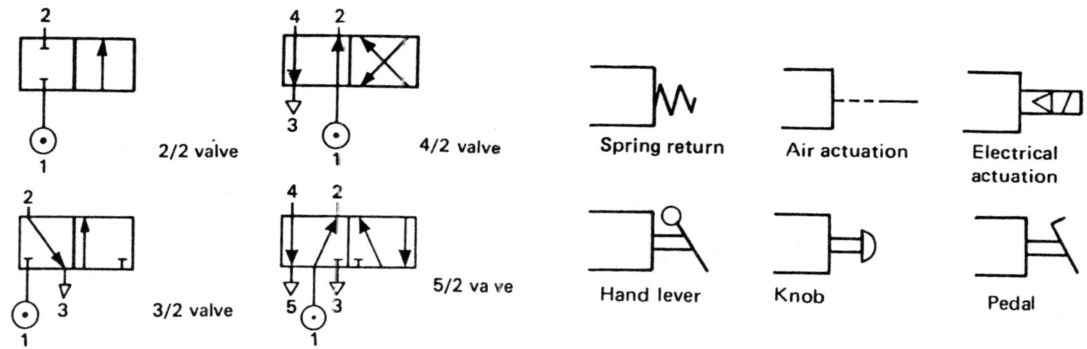

Figure 6.22 Standard pilot valve symbols.

by a solenoid operator. The solenoid operator is composed of a coil, plunger, and spring. The coil is a doughnut-shaped electromagnet. When voltage is applied to the coil power leads, the resulting current generates a flux field. This flux field acts like a magnet. A spring and plunger assembly is fitted through the core of the coil. The magnetic field of the coil pulls the plunger inward when the coil is energized, and the spring forces the plunger back to its normal, or original, position when the coil is deenergized. Other means of actuating a pilot valve are manually, by a push button or lever; electrically, by single or dual coils; and pneumatically, by a piston or diaphragm assembly.

Many pneumatic actuators are manufactured with an integral air manifold and internal porting that allow the pilot valve to be mounted in modular fashion directly onto the actuator (see Fig. 6.23). This eliminates the cumbersome adaptation of traditional solenoid valves with nipples and tubing, making a neater, more dependable package.

When selecting a solenoid-operated pilot valve for an actuator, it is important to specify that it be energized to open or energized to close. This means that the pilot valve should be ported in such a way that when voltage is applied to it (energized), it will pressurize the actuator, which in turn will operate the process valve to the open or closed position. With some integrally mounted solenoid valves, this se-

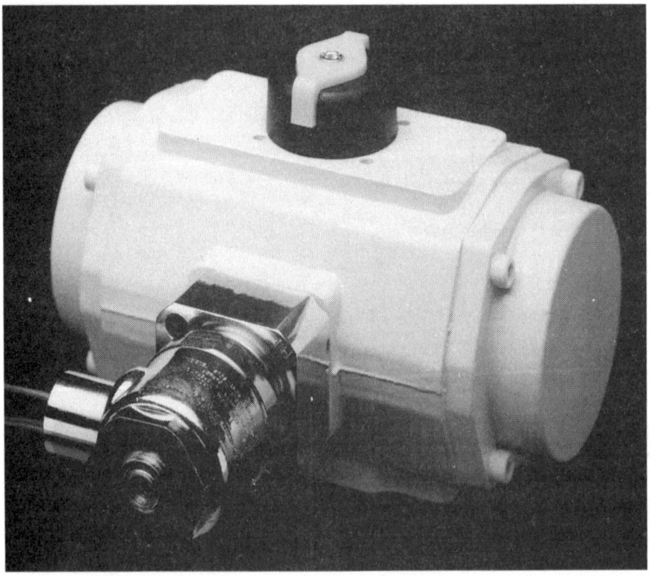

Figure 6.23 Direct-mount valve accepts electric single to pilot pneumatic actuator. (*Courtesy of El-o-matic, Inc.*)

lection is satisfied by mounting the valve in one position or 180° out of position.

To increase the speed of actuation a pilot valve with a higher flow coefficient may be used. In some cases the actuator may require modification to match the port size of a larger valve. For slower actuation a variable restrictor (needle valve) may be connected to the exhaust port. Undesirable speed and operation may result if the restriction is connected to the supply pressure port.

Limit switches

For a pneumatic actuator, the term *limit switch* may be a misnomer. The term more properly applies to *electric* rotary actuators that are fitted with limit switches to interrupt the power to the motor when the actuator has reached its desired limit of rotation. As a functional term, *position-indicating switch* is more properly applied to limit switches when they are used with pneumatic actuators (see Fig. 6.24). Indeed, a switch fitted to a pneumatic actuator does not limit its travel but instead indicates (through switches) when the actuator has reached, or has not reached, a specified point of rotation.

Also referred to as a switch box, the position-indicating switch encloses the switch elements, cams, and terminal strip and has a rotating input shaft that is fitted to the auxiliary shaft of the actuator to pick up rotary motion. The switch housing is composed of an input

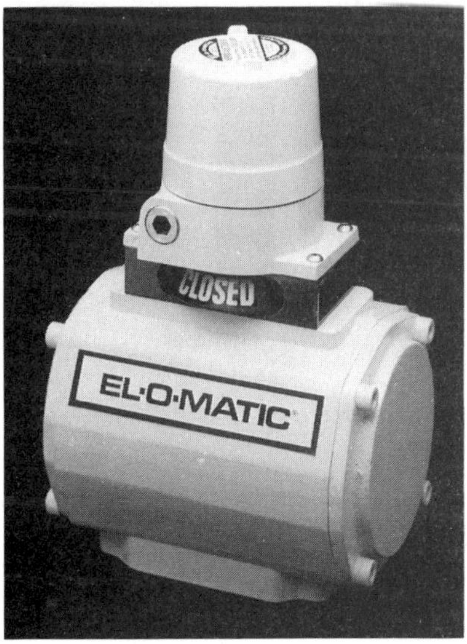

Figure 6.24 Top-mounted rotary limit switch on a pneumatic actuator. (*Courtesy of El-o-matic, Inc.*)

Figure 6.25 Inside view of switch enclosure shows switches, cams, and terminal strip. (*Courtesy of El-o-matic, Inc.*)

shaft that externally couples to the actuator's auxiliary drive shaft and is fitted internally with adjustable cams, snap-acting switches that are mounted to align with the cams, and a terminal strip for incoming wiring (see Fig. 6.25). As the actuator cycles, the input shaft of the switch box rotates and the cams actuate the switches. When the switches are used to indicate the limits of the cycle, the cams are adjusted to operate the switch when the desired position is reached.

Position-indicating switches are used for a variety of applications: light indication (powering indicator lamps on a control panel), system sequence cycling, alarms, electrical interlocking, etc. Some switch enclosures may be fitted with other devices, such as a potentiometer or position transmitter for continuous feedback of the valve's position.

When the switches are connected to signal lights, they should be arranged so that both lights are on in midtravel, with one or the other being extinguished at the ends of travel. This helps the operator avoid being misled by a burned-out lamp.

Switch boxes for pneumatic actuators are often specified by the type and quantity of switches required. Examples of the types of switches available are snap acting, proximity, and pneumatic switches.

Electric limit switch configurations

Electric switches are usually expressed in terms of the number of *poles* and *throws* they contain. A pole is a component of the switch that is moved by the switch action to make or break electric contact. The possible electric connections that can be made by a given pole are called throws.

For example, the most elementary switch configuration is shown in Fig. 6.26.

Figure 6.26 Single-pole, single-throw switch (SPST).

There are four configurations of electric limit switches: single-pole, single-throw; single-pole, double-throw; double-pole, single-throw; and double-pole, double-throw.

Single-pole, single-throw (SPST). This switch has one movable component, or a *single pole.* It also has only one possible connection, or a *single throw,* for that pole. The configuration of single-pole, single-throw is often abbreviated SPST. It is usually used as an on-off switch.

Single-pole, double-throw (SPDT). Figure 6.27 shows a single-pole, double-throw configuration. This switch still has only one movable component, but it has two possible connections, or throws. It is often found in a switch box where the customer may select between normally open or normally closed contacts. Normally open contacts provide an open circuit when the switch is in a free position, while normally closed contacts provide a closed circuit when the switch is in a free position.

Figure 6.27 Single-pole, double-throw switch (SPDT).

Double-pole, single-throw (DPST). This switch (Fig. 6.28) has two movable components, so it is a double pole. There is only one connection for each pole and, therefore, is a single throw. Note that the key

Figure 6.28 Double-pole, single-throw switch (DPST).

phrase in determining the number of throws is "for each pole." The contacts of this switch are normally closed.

Double-pole, double-throw (DPDT). This switch (Fig. 6.29) has two movable components, and, therefore, is a double pole. It also has two connections for each pole, so it is a double throw. This type of switch is often used in a switch box to give the customer an extra set of contacts to wire indicator lights or relays. Almost any switch configuration can be identified by applying the principles outlined here.

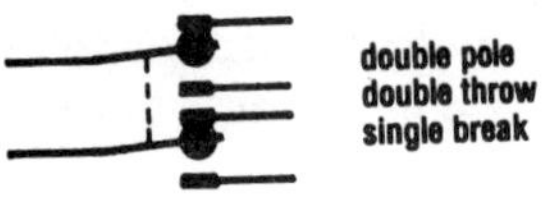

Figure 6.29 Double-pole, double-throw switch (DPDT).

Proximity switches

These switches operate when a metallic or magnetic object is brought into close proximity to the switch sensing area. These switches are inherently protected against dust and moisture and some require a power circuit. Two types of proximity switches are the proximity sensor and reed switch.

Proximity sensors. Proximity sensors are switches that operate when a metallic object is brought into close proximity to the sensing face. Most proximity sensors comply with several NEMA ratings. The sensors are protected against dust, moisture, and oil. Internal solid-state circuitry prevents shock and vibration from affecting sensor operation.

Reed switches. Another low-current proximity switch (250 to 500 mA) is the reed switch. Action is initiated when a magnet is placed in the proximity of the sensing area. Reed switches do not require a power supply.

Pneumatic switch

In some instances a pneumatic valve may be used to indicate actuator position. This method is sometimes used in hazardous areas to eliminate electrical equipment. A roller and cam are used to initiate valve action. At the control panel a pressure switch or pneumatic or mechanical sensor will indicate the presence of air pressure and indicate the valve position.

Other methods of position indication

If continuous monitoring of an actuator's position is required, as in modulating or "jogging" applications, a switch box may be fitted with a potentiometer. As the shaft of the switch box rotates, it likewise rotates the input shaft of the pot. The continuously decreasing or increasing resistive signal may then be converted into a valve position at the control panel. When the actuator is located far from the control system, the result may be an unreliable resistive signal due to the inherent resistance of the long wire. In this case a resistance-to-current transducer circuit may be preferred. The circuit board is usually installed in the switch box with the potentiometer and provides a 4- to 20-mA signal to indicate valve position.

Switch boxes designed for use in explosive environments must be able to withstand an internal explosion without igniting the explosive mixture surrounding the switch enclosure. The enclosure is thus designed to withstand the maximum expected internal explosion pressure without damage or excessive distortion and to provide venting for the pressure through channels of such dimensions that gases will be cooled below the ignition temperature before reaching the surrounding atmosphere. Thus the design of an explosion-proof switch enclosure involves careful consideration of housing thickness, cover fit, and tolerances.

Many switch enclosures incorporate multiple construction standards (NEMA I, IV, VII, IX, etc.) to satisfy a wide range of applications. (See Chap. 11 for an explanation of NEMA.)

Pneumatic positioners

When a valve is used for throttling rather than simple on-off service, it may be considered a rotary control valve. A control valve is a process control element that varies the flow of fluid as required by a process in response to a system control signal. To provide fast, sensitive, and accurate positioning in response to a control signal, an actuator must be fitted with a pneumatic positioner. A pneumatic positioner is basically a relay that senses and compares an instrument signal and the valve stem position. Because it is usually mounted to the top of a rotary actuator it actually senses valve position through the actuator shaft.

Most basic positioners have linear characterization. This means that the input signal to output rotation is directly proportional, which enables the process engineer to select a valve that will provide system characteristics. Standard ball valves, for example, provide equal percentage flow like many other quarter-turn valves do.

Operating principle of a positioner (3- to 15-lb/in^2 signal range). Referring to the schematic illustration (Fig. 6.30), with air pressure applied to the supply port of the pilot, and the instrument signal at midrange [9 lb/in^2 (0.6 bar)], the unit is in force balance. The instrument signal acts upon a diaphragm assembly creating a force toward the pilot valve. An opposing force from the suppression spring counteracts this force. The cantilever range spring is in its midzero position. An increase in the instrument signal causes an increased force toward the pilot valve and moves it open, increasing the output pressure to the actuator. This increase causes the actuator to rotate further and rotates the cam clockwise. A cam follower rises and twists the cantilever range spring in the counterclockwise direction resulting in a force at the end of the cantilever range spring directed away from the pilot valve. The actuator continues to rotate until the force of the cantilever range springs equals the increase in the instrument signal above its midpoint.

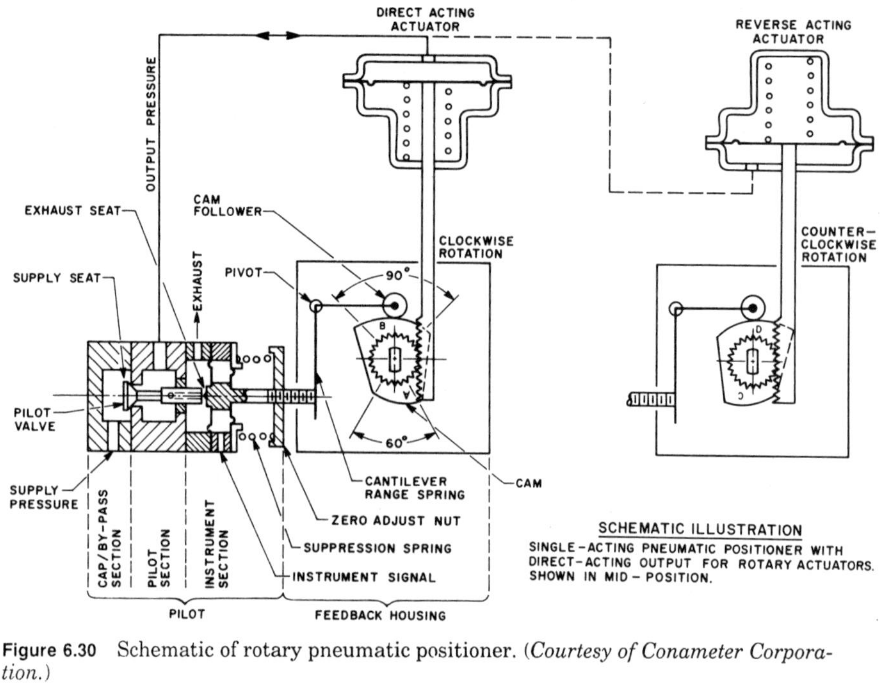

Figure 6.30 Schematic of rotary pneumatic positioner. (*Courtesy of Conameter Corporation.*)

A decrease in the instrument signal from its midpoint causes the diaphragm assembly to move away from the pilot valve and open the exhaust seat. As the output pressure decreases, the actuator rotates counterclockwise and rotates the cam counterclockwise. The cam follower drops, which twists the cantilever range spring in the clockwise direction, resulting in a force directed toward the pilot valve. The actuator continues to rotate until the force of the cantilever range spring equals the decrease in the instrument signal below its midpoint.

In a typical plant, the dynamic requirements of the instruments and valves with positioners represent an important factor in selecting these instruments, particularly the steady-state air consumption. The cost per year of one standard cubic foot of air bled or leaked to the atmosphere from a pressurized header may be estimated using the following formula:

$$\frac{\text{Dollars/year}}{\text{SCFM}} = \frac{\text{HP}}{100\ \text{SCFM}}(\$/\text{Kw} \cdot \text{hr})(75)$$

where SCFM is standard cubic feet per minute and HP/(100 SCFM) is the compressor factor.

The steady-state air consumption of pneumatic positioners depends on the type of pilot used. Positioner pilot valves are either spool valve types or one- or two-stage closed-center poppet types. One manufacturer uses an open-center valve type; another uses a mechanical-tilting flapper. All things being equal this formula indicates that factoring the steady-state air consumption of each positioner type being considered could indicate dramatic effects on the cost of operating such a system. For instance, a typical spool-valve pilot having an air consumption of 0.5 SCFM (15 L/m) at 80 lb/in^2 (6 bar) supply would cost twice as much to operate per year than a single-stage, closed-center poppet-type pilot.

Pneumatic positioners may be used with a variety of input control signals [3 to 15 lb/in^2, 6 to 30 lb/in^2, etc. (0.2 to 1 bar, 0.4 to 2 bar)]. To match positioner operation to these signals they are available with different range springs. This is useful when split-ranging.

Terms associated with positioners

Direct acting. Increasing input signal opens the valve (increases flow).

Reverse acting. Increasing input signal closes the valve (decreases flow).

Resolution. The smallest possible change in valve position.

Deadband. The maximum range through which the input signal can be varied without initiating change in valve position.

Hysteresis. The maximum difference in valve position for a given input signal during a full range traverse in each direction.

Transducers. A transducer is a device that converts one signal type to another. In the case of control instrumentation, a current to pneumatic transducer accepts an analog milliamp control signal from a field instrument and converts it to a proportional pneumatic signal for the positioner. The most common conversions used with control valves are

Input	Output, lb/in^2
4 to 20 mA	3–15, 3–9, 9–12
0 to 10 V dc	6–30, 6–18, 18–30

For systems being controlled and monitored with electronic instrumentation but with pneumatically actuated control valves, transducing is the most practical method for interfacing the two types of equipment. As an electromechanical device, a transducer must be carefully selected for environmental compatibility, hazardous areas, sensitivity, vibrations, etc.

One drawback of transducers is that it is sometimes difficult to locate them near the positioner, which may then require long runs of wire or pneumatic tubing. To satisfy this, some manufacturers have integrated the transducer into the positioner. These hybrids are known as electropneumatic positioners.

Standard instrument signals. Instrument signals are used to interface between various elements in the control process. Information may be transmitted from a sensor to a controller, or a controller to an actuator, etc. Standard instrument signals allow a wide variety of products made by different manufacturers to work together. Common standard instrument signal ranges are shown below. The high end of a standard instrument signal range is usually 5 times the value of the low end. For instance, 20 mA is 5 × 4 mA, 15 lb/in^2 is 5 × 3 lb/in^2, etc.

The low end usually does not have a value of zero. This provides a positive method of determining the difference between a device that is indicating the low end of a range and a device that is not functioning. This is known as *live zero*.

The main exceptions to these conventions are resistance-type inputs which usually have a low end of zero and various values of high ends. Split ranges are usually fractions of standard instrument signals. For example, 3 to 15 lb/in^2 is often split into 3 to 9 lb/in^2 and 9 to 15 lb/in^2, each of which is half of the standard range.

In pneumatic devices, pressure [lb/in^2 (bar)] is the usual variable for instrument signals. In electric devices, the variable may be current (mA), dc voltage (V dc), or resistance [ohms (Ω)]. The following table gives instrument signal ranges for pneumatic and electric devices.

Instrument Signal Ranges

Range type	Pneumatic (pressure), lb/in^2	Electric		
		Current, mA	DC voltage, V dc	Resistance, Ω
Standard	3–15	4–20 10–50	1–5 1–10	0–10,000 0–5,000 0–135
Split	3–9 9–15	4–12 12–20 10–30 30–50	1–3 3–5	

Uses of a pneumatic positioner. The following are some uses of a pneumatic positioner.

1. Temperature control
2. Level control
3. Split ranging
4. Loops with slow response
5. Reverse action relative to actuator

Split ranging is a process by which the input signal range [3 to 15 lb/in^2 (0.2 to 1 bar)] is used to pilot two control valves (see Fig. 6.31). In practice, the first control valve cycles through its full stroke in the range 3 to 9 lb/in^2 (0.2 to 0.6 bar), and the second valve strokes through the 9- to 15-lb/in^2 (0.6- to 1-bar) range.

Positioners are available in a variety of materials of construction, accessories, characterized cams, position transmitters, and integral transducers.

Manual override devices for pneumatic valve actuators. In this era of automation, it is possible to have more control over a process system than ever before. In fact, when the decision is made to automate a valve for a specific function or functions in a system, one important

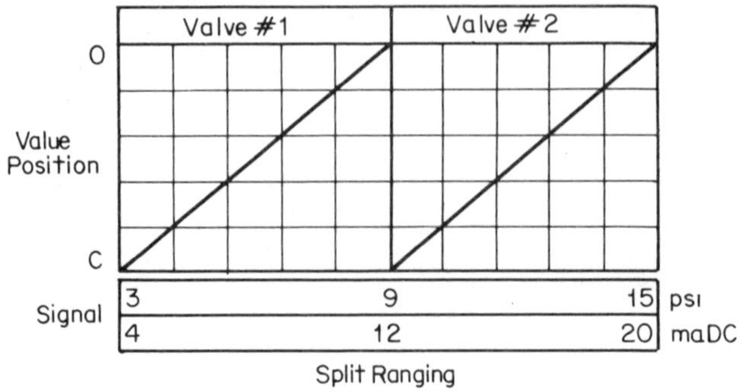

Figure 6.31 Split ranging is a technique by which the input signal range is used to pilot two control valves.

reason is to have even more complete control over the process by providing feedback, sequencing, and rapid response and by eliminating human error.

Interfacing an automated valve with a control system may require that an actuator be equipped with a solenoid-actuated pilot valve, positioner, limit switches, a mechanical position indicator, transducer, and so many other control accessories that if there is a loss of power to the actuator, or if the actuator fails to operate for any reason, it may be rendered inoperable and therefore useless, becoming a potential hazard or causing an unnecessary shutdown of the production process. In effect, control of the valve and possibly the entire process may be lost.

The simplest and most reliable method for guaranteeing the continued operability of an automated valve in the event of a system failure is to use a manual override device. As more quarter-turn valves are being incorporated into expanded process control systems, there is an increased concern over the ability to operate these traditionally manual valves in the event of actuator or power failure. This concern has been recognized and addressed by actuator manufacturers. There are currently a number of manual override provisions available for pneumatic quarter-turn actuators. We discuss four of them here: wrench override, hydraulic pump, jack screw override, and disengageable gear manual override.

Wrench override. A wrench override is simply a handle with an engagement provision that fits to the auxiliary drive shaft of the actuator (see Fig. 6.32). Upon failure, the wrench may be applied to the flats of the shaft to manually override the actuator. This method should be used only with double-acting actuators as it is difficult to override and hold spring-return actuators in position. Torque should be limited to about 1500 lb · in (170 N · m).

The wrench is usually attached to the actuator or mounting bracket

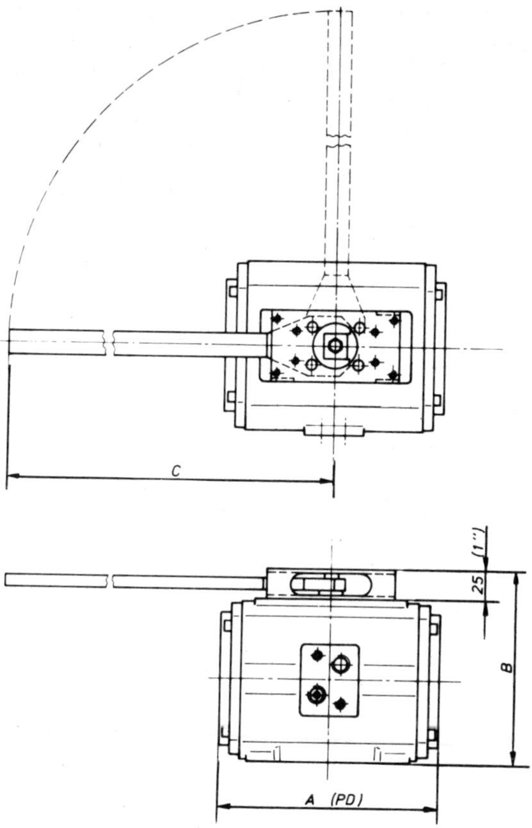

Figure 6.32 Wrench override may be used with small automated valves. (*Courtesy of El-o-matic, Inc.*)

with a cable or chain to prevent loss. It may also be available with a locking provision to hold smaller spring-return actuators in position until the problem is resolved. A wrench override should never be permanently attached to the drive shaft of the actuator because, when it operates automatically, it may cause injury to personnel working near the equipment.

Hydraulic pump. The hydraulic pump is a manual override system that is used with extremely high torque actuators that are fitted with an auxiliary hydraulic cylinder (see Fig. 6.33). Upon air failure, operating the crank input on the pump generates hydraulic pressure, which is applied to the high-pressure cylinder, thereby overriding the actuator.

Jack screw override. A jack screw override is a length of heavy-duty threaded stock that is tapped into the end caps, or canisters, of cylinder or diaphragm actuators (see Fig. 6.34). Rotating the jack screw,

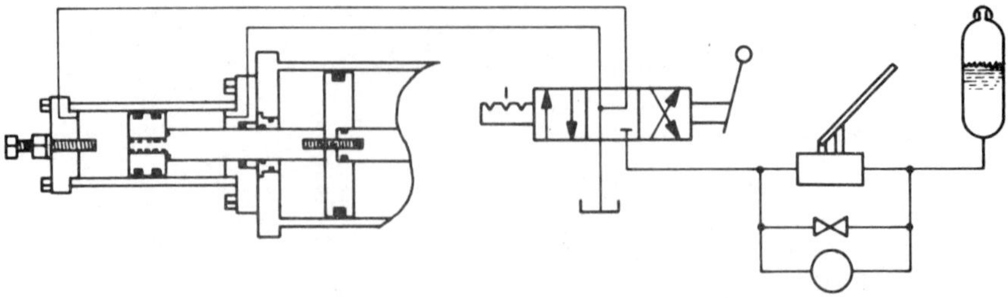

Figure 6.33 A hydraulic pump manual override provides auxiliary power in the event of air failure.

which is usually fitted with a handwheel, creates a linear force that is applied to the piston face, thereby overriding the actuator. For spring-return actuators, only one jack screw is required. The actuator position is held until the screw is backed off. Spring action drives the actuator to its original position.

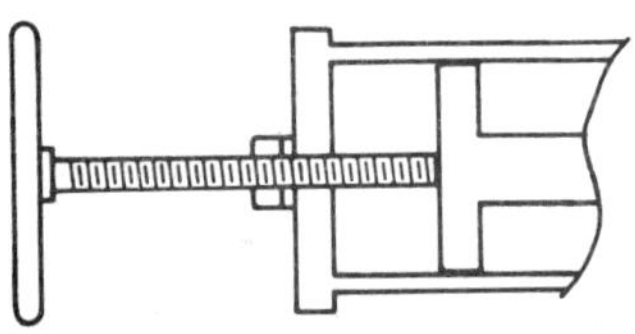

Figure 6.34 Jack screw overrides are the most economical means of overriding an actuator.

For double-acting actuators, two jack screws are required, one for each direction of travel. Because they are installed on the ends of the actuator, jack screws add considerably to the overall length of the assembly and may be prohibitive in tight spaces.

Jack screws must also be protected from damage by weather and corrosion, which could cause difficult operation and eventual seal failure. Finally, if the jack screw is bent, it becomes virtually inoperable, defeating the purpose of this safety device.

Disengageable gear manual override. The disengageable gear override is a modular component that fits between the valve and the actuator and offers simple, reliable manual positioning (see Fig. 6.35). The self-locking bronze-on-steel worm gear design provides for safe and easy operation and positive manual positioning even with spring-return actuators. Rotating the clutch lever, located at the base of the handwheel, 180° immediately engages the worm gear with the output drive sleeve to permit operation. Manual override modules may be adapted in the field to existing control valves with a slight modification to the actuator.

Figure 6.35 Manual gear override is used with small and large automated valves. (*Courtesy of El-o-matic, Inc.*)

Manual overrides have proven to be an accessory requiring greater consideration in many applications. Modular construction, immediate operation, and adaptability to standard actuators are important to consider.

Chapter

7

Electric Actuators

7.1 Electric Actuators for Rotary and Multiturn Valves

Electric actuators are used for the automation of process valves when other available means of actuation (pneumatic or hydraulic) are either unavailable or impractical. They are used extensively in industries like power and in the petrochemical market to power large gate valves and in the water and waste market for the automation of butterfly valves. Like the valves that they are intended to automate, electric actuators come in two major styles: multiturn and quarter-turn, sometimes called linear (multiturn) and rotary (part-turn). Generally speaking all electric actuators are geared electric motors with position-sensing elements (limit switches) and/or torque-sensing elements (torque switches) that shut off power to the motor when the desired position or amount of torque has been achieved. Because they are an electric device anyone who uses an electric actuator must consider (1) the area classification and (2) the severity of moisture and temperature in the operating environment when specifying electric actuators.

Linear and rotary electric actuators have many things in common regarding control accessory options, and except for some manufacturers' philosophies on gearing systems and other performance elements, the following discussion of electric actuators applies fundamentally to both types of actuators.

Understanding electric actuators for valve automation

Though the pneumatic actuator will probably continue to be the workhorse throughout the process industry for the automation of rotary valves (ball, plug, and butterfly valves), there are many applications where an electric actuator may be considered. For the process control engineer a good understanding of the basic function and features of electric actuators and their options and versatility will lead to a better technical and practical evaluation of their usefulness.

Why use electric actuators? Although electric actuators may be used anywhere a power source (electricity) is available, there are many applications where they are particularly well suited. For instance, in many remote installations it may be impractical to run an air supply and to maintain it. Air lines that freeze up may clog and render the equipment inoperable to damage more delicate instruments. If only a few actuators are to be installed in an area, electric actuators offer a simple means of automation for these smaller systems.

Computer compatible. Perhaps one of the most important reasons for the trend toward using electric actuators has been the decreasing cost of using computers as system controllers and the ease and economy with which the actuators can be interfaced to such systems. This trend can be expected to accelerate with the advent of new microprocessors, or smart controllers, based on the versatile, low-cost microprocessor chips. Because of the increased speed and decision- and control-making capability that the computer adds to a process system, there is less need for final control elements which have high control capability such as characterized globe and plug valves. As a result, the simpler and less expensive electric actuators and ball and butterfly valves have become more acceptable and are proving more than adequate for many applications.

How to evaluate costs. Generally, it is more economical to install and maintain electric valve actuators than pneumatic ones. A pneumatic system includes not only the actuators but compressors, piping, filters, air lubrication systems, and dryers. After installations, its many components cost more to maintain and usually require the coordinated service of both the plant's electrical and plumbing departments. Where electronic sensors are used with pneumatically controlled systems, the cost of current-to-pneumatic (I-P) converters must be added if positioning of the actuators is required.

Electric actuators eliminate the need for air, an expensive source of energy, and do not require energy when not in motion. There is no need to be concerned with compressor noise, housing to shield against noise, air venting, or other operating restrictions associated with pneumatic systems.

How electric actuators work

An electric actuator is basically a geared motor. The motor is the primary torque-generating component. Motors are available for a variety of supply voltages including standard single-phase 110 V ac, three-phase, and dc voltages. To prevent heat damage due to excessive current draw in a stalled condition, or due to overwork, electric actuator motors usually include a thermal overload sensor embedded in the winding of the stator. The sensor is installed in series with the power source and opens the circuit when the motor is overheated and then closes the circuit once it has cooled to a safe operating temperature.

Why motors fail. If a motor operates at higher than allowable temperatures, the excessive heat will degrade the insulation and shorten the motor's life. In fact, if two motors are operating at a 50°F (10°C) differential, the motor with the higher temperature will have only half the life of the cooler motor.

The NEMA has established classes of insulation systems and allowable operating temperatures for each. Classes B and F are the most commonly used. The maximum ambient operating temperature for both classes is 104°F (40°C). The maximum allowable temperature rise varies between classes and a motor's service factor. For example, a class B design with a 1.0 service factor has a total allowable temperature rise of 270°F (130°C). A class F design with a 1.15 service factor has a total allowable temperature rise of 310°F (155°C). Therefore, a motor using class F insulation, with its higher thermal capacity, in a class B motor, with its lower thermal capacity, will actually operate below its temperature rating. It will last longer.

Consider the following causes of motor failure.

Overload usually results from some change in the driven load, a broken gear, or frozen bearing. If motor-running protection devices take the motor off the line repeatedly, the cause should be located. If a motor has a 1.15 service factor, do not run the motor at its 15 percent "cushion" factor continuously. This is for temporary overloads.

Short cycling is repeatedly starting a motor which has already reached operating temperature. Motor starting current can be as much as 5 to 6 times full load running current. Most electric actuator manufacturers specify the starting frequency for their products.

Moisture in the form of condensation can cause rust within a totally enclosed motor. As a motor cools down, air with its moisture is sucked inside the motor. Good motor designs, mounted externally, have a small drain hole in the bottom of the unit to allow moisture and air to escape. If motors are stored in a moist atmosphere, the insulation, even though it is nonhygroscopic, may lose some of its insulating properties. Motors should be stored in a warm, dry place. Flooded motors can usually be put back into service after a good baking. Electric heaters can be installed in large motors to drive out moisture. In electric actuators, compartment heaters provide for a constant operating temperature and eliminating cooling and breathing.

Low or high voltage is not the cause of failure as frequently as suspected. Good electric systems and operating practice limit the voltage at the motor under load to not more than ±5 percent of nameplate. This is well within limits of NEMA specifications, but today's motors do not give satisfactory service if the voltage is consistently about 5 percent above nameplate rating.

Duty cycle. Electric actuators are most commonly available with a 25 percent duty cycle design. This means that to prevent overheating the motor, the unit must be turned off for three cycle periods before being operated again. With optional motors and/or capacitors, or by sizing the actuator for valves with lower torque requirements, an electric actuator may be upgraded to 100 percent duty cycle. This is also referred to as continuous duty.

Most often a 100 percent duty cycle motor is specified in control applications. Here, continuous modulation of the valve position results in excessive motor starts per unit of time. Most control valves however remain in a given position for long periods of time. Especially with today's new electronic positioners constant "hunting" and overshooting are for the most part less troublesome for electric motors. If positioners with slow-approach features are used, then motors with a 50 percent duty cycle are usually acceptable in modulating service.

How duty cycle is calculated. Motor manufacturers calculate the duty cycle by operating the motor continuously until it overheats. After overheating, the motor is allowed to cool and recover. When the motor has cooled sufficiently, it resumes operation. The total length of time for operation and recovery is recorded. Following the test, the recorded length of time is evaluated and broken down into two segments. Those segments are (1) percent of running time and (2) percent of recovery time. These two percentages are used to establish the duty cycle rating.

$$\text{Percent duty cycle} = \frac{\text{motor running time}}{\text{motor running time} + \text{motor cooling time}} \times 100\%$$

Several factors affect the duty cycle of a particular motor. One is the temperature of the surrounding atmosphere, often referred to as the *ambient temperature*. Most duty cycles are determined at room temperature, or 72°F (20°C). A motor that had a 20 percent duty cycle at room temperature would have to be derated considerably if installed in a high-temperature environment. Likewise, if a motor is installed in an area where the temperature is lower than 72°F (20°C), the duty cycle may be increased.

Another method for establishing a duty rating for an electric motor is derived from the number of starts expected over short periods of time. This is particularly important to consider when an actuator is used in modulating service. Continuous hunting for the desired valve position requires the motor to stop and start as it seeks the signal position. Starts-per-hour ratings are a function of the duty cycle rating of the motor and ambient temperature. The following table is an example of this rating.

Ambient temperature			
°F	°C	Duty cycle, %	Starts per hour
-22 to 104	-30 to 40	100	1200
104 to 140	40 to 60	100	630
140 to 158	60 to 70	100	516
-22 to 104	-30 to 40	50	600
104 to 140	40 to 60	50	315
140 to 158	60 to 70	50	258

Limit switches. As an electric device, the action of a motor is initiated by the introduction of voltage. Conversely, cutting off the power to a motor will, of course, shut it off. To precisely set the rotary output position of an electric actuator a system of cams and limit switches is used. In most designs a cam shaft is connected to the final output gear of the actuator gear train. The shaft is fitted with two adjustable cams each acting upon a microswitch. One switch is used to cut off power to the motor in the open position and the other is used for the closed position. Limit switches should always be calibrated while the actuator is on the valve to ensure that proper valve closure is achieved.

Understanding switch terminology. Microswitches are snap-acting electric devices used to permit or prevent (on or off) the flow of current in an electric system. The most commonly used microswitch in electric rotary actuators is known as a single-pole, double-throw (SPDT) switch. SPDT switches have three terminals and are labeled normally open, normally closed, and common.

Normally open. For any switch the *normal* condition is that which the device takes when there are no external forces (like a cam) acting upon it. The *open* condition means that the switch circuit is open with no current flow.

Normally closed. In this normal condition the switch contacts are made and the flow of current is permitted.

Common. The normally open and normally closed switch contacts require some point to make contact with. Because they share a third point (which is usually the source of incoming power), it is usually referred to as the *common.*

Cam design. There are two basic cam designs used with electric rotary actuators—eccentric and flatted. Eccentric cams act upon a limit switch to shut off power to the motor. This is achieved when a high point on the lobe of the cam acts upon the plunger of the switch and pushes it in to cut off power. These switches carry motor current through their normally closed contact. Flatted cams are designed so that they are in constant contact with the switch (closing the normally open switch contact) and interrupt power to the motor when the switch plunger reaches the flatted surface of the cam. The advantage of the flatted cam is that the normally open contact of the switch carries the motor voltage which will not pass current if the cams have not been installed. This prevents unwanted cycling of an actuator that has not been properly set.

Typical control circuit. A typical electric control circuit for a single-phase actuator includes a motor, two cams and two limit switches, light wiring, customer wiring, and terminal points. Using the circuit in Fig. 7.1 we will follow the path of current from initiation to shutoff for an electric actuator. This exercise is important in demystifying the function of electric actuators and in considering optional circuit designs for more unique applications.

When the selector switch is moved to the open position, voltage is applied to the actuator's motor through the normally closed contact of the closed limit switch. The motor torque is applied to and enhanced through the gear train which in turn rotates the process valve in the opening direction and at the same time rotates the cam shaft. The cam position has been preset so that when the valve reaches the desired full-closed position, the limit switch plunger is actuated by the cam. This cuts off the supply voltage to the motor and, through the normally open switch contact, energizes the closed position indicating light at a remote control panel.

A three-phase motor is energized through a reversing starter. As an electromechanical device it uses single-phase power to energize the

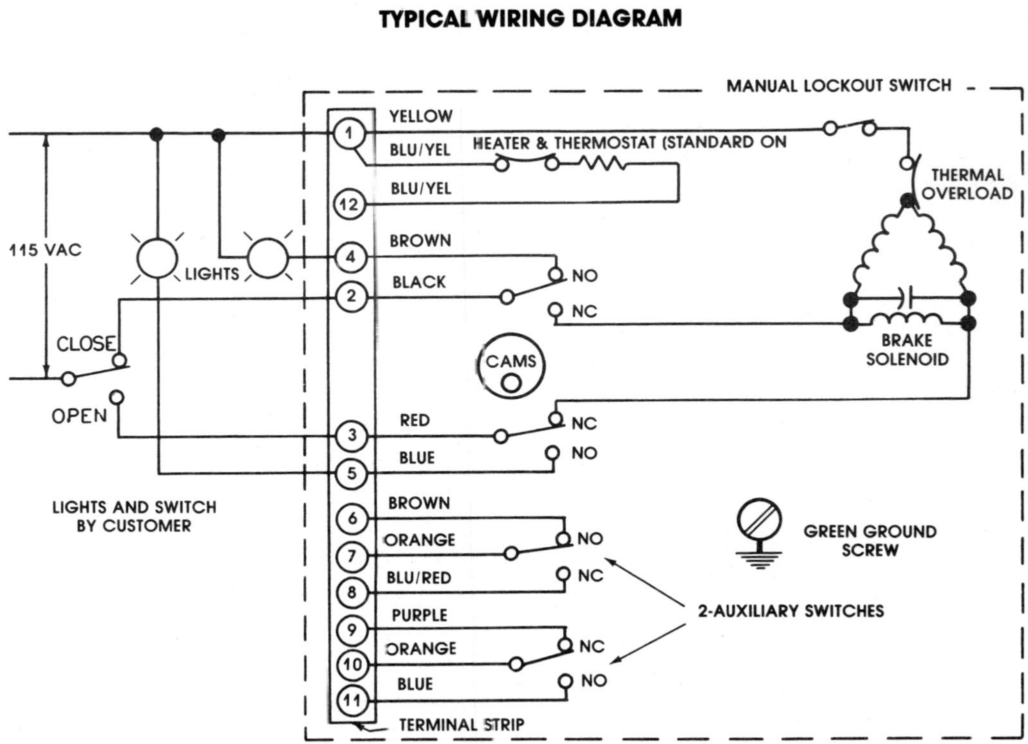

Figure 7.1 Typical single-phase wiring diagram for reversible electric actuator. (*Courtesy of Demco Division of Cooper Industries.*)

coils that cause mechanical action of the contacts. The limit switch is wired in series with one side of the coil for each direction of rotation.

Control accessary options

Probably the most important reason for the widespread use of electric actuators is their control circuit versatility. As an electric device, electric actuators naturally lend themselves for use as an enclosure for a variety of control and feedback devices. Furthermore, the switch and cams may be set and wired for almost any contact development for process and valve control.

A few examples of the control circuit versatility of electric actuators are three-position control and interposing relay. Three-position control is especially useful for the automation of multiported (three- or four-way) valves. An interposing relay is powered by a simple on-off switch. This circuit is similar to the type used in energizing single-acting solenoid valves in pneumatic actuator installations.

One advantage of an electric actuator is its ability to manually "jog" the valve position when it is used in filling operations. By installing a feedback potentiometer an operator can monitor the actuator's exact position and stop it at any point between open and closed with a manual control switch.

Gearing. Electric actuators utilize a gear train (a series of interconnected gears) to enhance the motor torque and to regulate the output speed of the actuator. Some gear styles are inherently self-locking. This is particularly important in the automation of butterfly valves or when an electric actuator is used in modulating control applications. In these situations seat and disc contact or fluid velocity act upon the closure element of the valve and cause a reverse force that can reverse the motor and cam shaft. This causes a reenergization of the motor through the limit switch when the cam position is changed. This undesirable cycling will continue to occur, unless a motor brake is installed, and usually leads to an overheated motor. Spur gears are sometimes used in rotary electric actuators but are not self-locking. They require the addition of an electromechanical motor brake for these applications.

A few of the self-locking gear styles include the worm and wheel and some configurations of planetary gears. A basic worm gear system operates as follows (see Fig. 7.2). A motor applies a force through the primary worm gear to the worm wheel. This in turn rotates the secondary worm gear which applies a force to the larger radius of the secondary worm wheel to increase the torque.

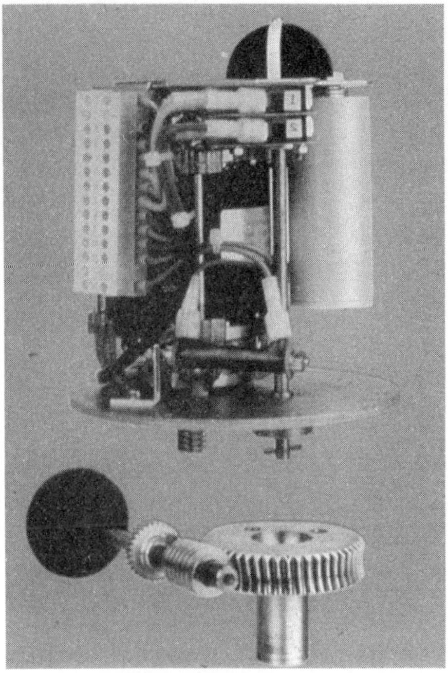

Figure 7.2 Exploded view of electric rotary actuator with double-reduction worm gear system. (*Courtesy of El-o-matic, Inc.*)

Manual override. Depending on the manufacturer, a manual override provision is sometimes an option. A manual override is important to consider when there is the possibility that a power failure or other malfunction may render a critical process valve inoperable (see Fig. 7.3). Because most mechanical overrides on electric actuators make use of the mechanical advantage of the gearing, it is possible, because of the potentially excessive force that may be applied to the input handwheel, to damage the gearing or the valve when an override is used. For this reason the handwheel diameter is selected based on a maximum rim pull of about 50 to 60 lb. With this input the resultant forces on the actuator gears (as well as its final output) will be more like that for which the actuator has been designed. This helps eliminate gear and/or valve damage.

Electric actuators are available with a wide selection of control accessories to meet many process control applications. Among the most widely offered are extra switches, torque protection, and mechanical brakes.

Extra switches. Additional switches may be added to most electric actuators. The quantity is generally limited by the physical space

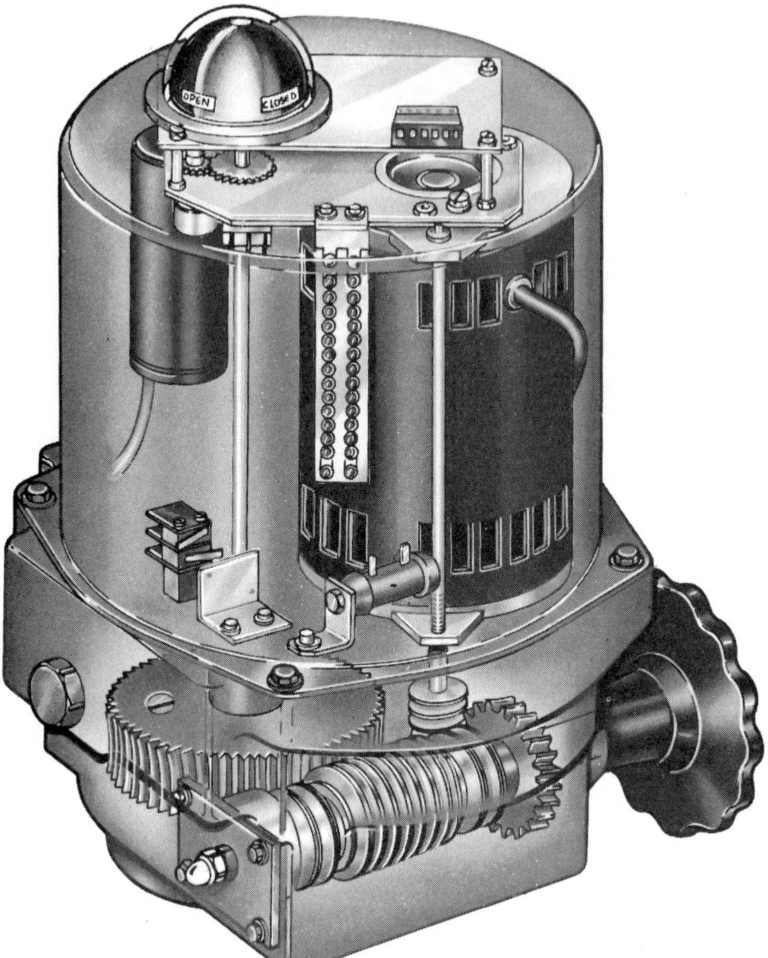

Figure 7.3 Electric rotary actuator with manual override. (*Courtesy of El-o-matic, Inc.*)

available. Extra switches may be used for intermediate position indication, sequencing, and system interlocking. Some control configurations, like three-position control, require additional switches.

Torque protection. Electric actuators are, of course, torque-generating devices. When the torque required at the valve stem exceeds the capacity of the actuator, the motor will stall. If voltage is not interrupted, the motor will continue to draw current and generate heat that will eventually damage it. In addition, applying torque to a valve that is jammed could twist or shear the stem. To prevent this condition from causing damage to the valve or actuator, some manufacturers offer a torque-protection package. The two most common types of torque protection are current sensing and motion (or torque) sensing. Current-sensing devices are solid-state current- (or heat-) actuated switches. Some are adjustable and will trip when a stalled motor begins to draw excessive current. Motion-sensing torque protection devices for worm-gear-style electric actuators sense a stall condition at the final output gear. In this condition the secondary worm gear will tend to "climb" along the stalled worm wheel. Worm gear motion is restricted by a belleville washer system installed at each end of the gear. A mechanical switch is driven by the linear motion of the worm gear. Its trip point is usually factory adjusted.

Multiturn electric actuators often use torque switches to shut off their motor when a gate or globe valve is properly seated. These thrust-seated valves may not shut off completely if the closure element is merely positioned by limit switches. To ensure adequate closing force these actuators are generally oversized and utilize fully adjustable torque switches.

Mechanical brake. As discussed earlier spur gear trains are not self-locking. For these designs an electromechanical motor brake must be used for the automation of butterfly valves in high-velocity or modulating control applications. Traditional electromechanical brakes use a solenoid to activate a rubber "shoe" that acts upon a brake wheel. The spring-loaded shoe rests on the wheel when the motor and solenoid are deenergized.

Potentiometer. A potentiometer is a variable resistor whose output is controlled by the rotation of its shaft. Radio volume knobs drive a potentiometer. The traditional potentiometer used with electric actuators is a wire-wound type, although other designs are also used. The potentiometer (also called a pot or potmeter) has three leads (see Fig. 7.4). The leads are connected to the ends of the wound wire, and if connected to a meter, will indicate the total resistance of the pot. The

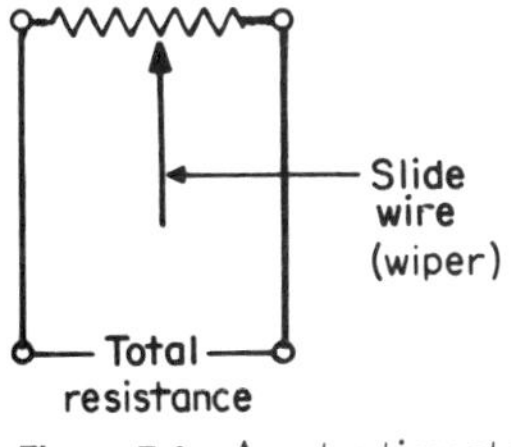

Figure 7.4 A potentiometer (variable resistor) is used to continuously indicate actuator position.

third lead, called a *wiper,* contacts the wire-wound resistor at various points depending upon the position of the rotary shaft. When a meter is connected to one of the end leads and the wiper, it will read a portion of the total resistance of the pot. The resistive value may then be translated to indicate the rotary position of the shaft. This characteristic of the potentiometer makes it ideal for use with rotary actuators where a continuous indication (a readout) of its position is desired.

Potentiometers usually have the following ratings:

1. Resistance range (such as 5 kΩ). The maximum resistance of the potentiometer.
2. Resistance tolerance (such as +20 percent). The maximum deviation from the rated resistance.
3. Power rating (such as ¾ W at 70°C). The maximum power that may be applied to the potentiometer without damage.
4. Voltage rating (such as 350 V dc). The maximum voltage that may be applied to the potentiometer without damage.
5. Angle of rotation (such as 270°). The shaft rotation that will cause the resistance to vary from minimum to maximum. Some potentiometers are multiturn and therefore have several complete turns of rotation.
6. Stop strength [such as 3 lbf · in (0.3 N · m)]. The maximum torque that may be applied without damage to the potentiometer shaft at the end of rotation.

A potentiometer (variable resistor) is used with electric actuators for three possible functions: (1) continuous resistive feedback of valve position to a control panel, (2) as a feedback of valve position to a milliamp position transmitter, and (3) as an input to the comparator circuit of an electronic positioner used in proportional control applications. If more than one function is desired, a dual-potentiometer may be installed.

Milliamp position transmitter. When it is necessary to monitor the position of an electric actuator, a feedback potentiometer may be used. For transmitting over great distances or for certain control systems, however, a continuous milliamp signal is often desirable. A milliamp position transmitter provides a feedback signal range of 4 to 20 mA over 90° of actuator output.

Proportional controller. Also referred to as an electronic positioner or servo amplifier, the proportional controller accepts a 4- to 20-mA input signal from a process controller. At the heart of the proportional controller is a comparator circuit that compares the desired position, indicated by the 4- to 20-mA input signal to the achieved position signal provided by the potentiometer. Because an electric actuator may be required to continuously make changes to the valve position, an extended duty motor is recommended (see previous discussion on duty cycle).

Operation of electronic positioner. The command signal for an electronic positioner (see Fig. 7.5) is often 4 to 20 mA dc but could be a number of other values. It tells the positioner where the actuator should be positioned. The actuator is usually designed to go fully closed at 4 mA and fully open at 20 mA when the positioner is set for direct-acting. When this convention is reversed (4 mA open, 20 mA closed), it is known as reverse-acting.

The actuator feedback signal often comes from a potentiometer or other resistive element. It is an electric signal that tells the servo where the actuator actually is positioned. The comparator circuit compares the command signal and the actuator feedback signal.

1. If the signals are different, the comparator activates either the open or closed output circuit to reduce the difference.

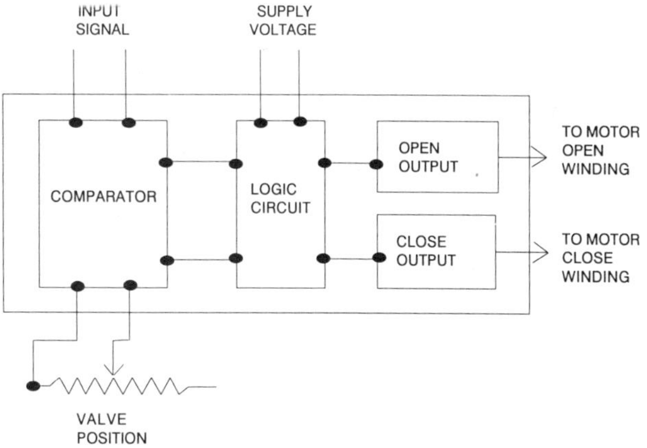

Typical electronic positioner circuit for electric actuator

Figure 7.5 Block diagram of electronic positioner used with electric actuator.

2. If the signals are identical, the comparator activates neither circuit.
3. The comparator should never activate both output circuits at the same time.

The output circuit features some form of switching element to connect the high power supply to the motor leads. The switching elements could be electro-mechanical relays or solid-state relays called TRIACs.

Deadband. The positioner requires a certain minimum difference between the command signal and actuator feedback signal in order for the comparator to activate an output relay. Suppose the actuator was at exactly 45° and the command signal was 12.1 mA. Most positioners would say that the actuator position was close enough to the command signal and would therefore not activate the motor. Further suppose the command signal were slowly increased from 12.1 mA and that the actuator began to move when the command signal was 12.3 mA. Now the difference is large enough to activate the motor. Between 12.1 and 12.3 mA, however, the actuator was dead. The term *deadband* defines the minimum signal change that must be given to the positioner in order to initiate action. It is usually expressed as a percentage of total command signal span.

A small deadband allows for greater accuracy in positioning the actuator. However, if the deadband is too small, the actuator will have difficulty reaching a balanced condition. This is often called *hunting,* since the actuator will oscillate between two points but never settle down.

Electronic comparators are very sensitive and are capable of detecting extremely small differences between signals. Most actuator deadbands are therefore limited by the capability of the motor and power gear train. It is not uncommon to find a 0.1 percent electronic device driving a motor and gear train that can only position itself within 2 percent. The deadband of the complete unit will only be as good as that of the poorest part.

Electronic speed control circuit. As a geared motor, an electric actuator will provide only one cycle speed from 0 to 90°. Some manufacturers offer a variety of geared speeds for specific needs, but for simplicity and availability a cycle length controller pulses power to the motor for a given period of time (generally 0.5 s), called the *on* time, and then shuts off for another period of time, called the *off* time, and then repeats the cycle. The off time is adjustable between 0 and 10 s which means that an electric actuator with a given cycle speed of 5 s for 90° can be adjusted to take as long as 95 s for a quarter turn of rotation.

To calculate the off period setting for a given cycle length use the following formula:

$$\frac{C_L - S_A}{S_A/\text{ON} - 1} = \text{OFF}$$

where C_L = actuator cycle length desired, s
S_A = actuator speed, s
ON = fixed ON time of cycle length circuit

Note: This formula is for speed control circuits that provide an immediate on period when the actuator is energized.

Example Let's say that process parameters require that an actuator cycles from open to closed in 50 s. The actuator has a given cycle speed of 5 s over 90° and the on period of the cycle length control circuit is 0.5 s. Now we have C_L = 50 s, S_A = 5 s, and ON = 0.5 s. Then

$$\text{OFF period} = \frac{50 - 5}{5/0.5 - 1} = \frac{45}{9} = 5 \text{ s}$$

Three-phase electric motor actuators and accessories

The availability of three-phase alternating current makes possible the simplest, cheapest, and most reliable form of the reversible electric motor—the squirrel-cage motor. A two-pole motor has one pair of poles and for 60 cycles per second [hertz (Hz)] runs at 60 revolutions per second (r/s) or 3600 r/min. A four-pole machine has a synchronous speed of 1800 r/min and so on. The corresponding speeds for 50 cycles/s are obviously 3000 and 1500 r/min, respectively.

Reversing operation. If two of the wires between the power source and the motor are reversed, the magnetic field on the motor will rotate in the opposite direction to that of the power generator. It is, therefore, only necessary to reverse any two phases to change the rotary direction of a three-phase motor, and this is the function of reversing contactors.

Contactors and relays. There is no difference in operating principle between a contactor and a relay. The term *relay* is generally used when the function of a device is to repeat a signal into another circuit or circuits; the term *contactor* is used when its function is to switch a

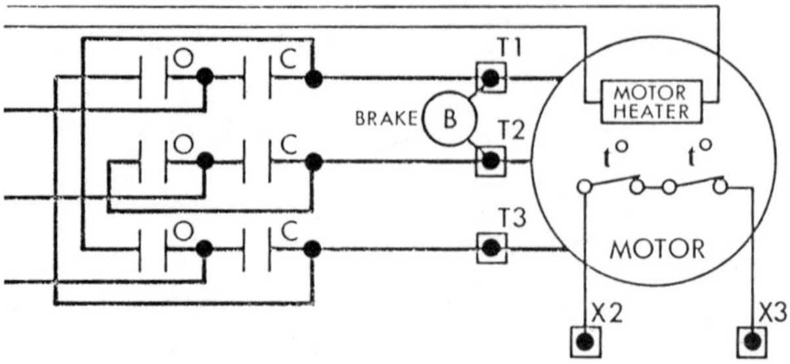

Figure 7.6 Reversing contactor uses low voltage (control voltage) to operate three-phase motors. (*Courtesy of EIM Controls.*)

load on or off. In either case, the device contains an electromagnet with a moving armature which has contacts to make or break a circuit when the magnet coil is energized. When the coil is deenergized, the armature is returned to a normal position by a spring.

A reversing contactor allows the use of low voltage, usually single-phase alternating current, as the control power for much higher powered three-phase motors (see Fig. 7.6). Contactor coils draw only a fraction of an amp when reversing the operation of electric actuators that may draw as much as 100 A.

Typical installations of reversing contactors use two coils, each activating a set of contacts. When deenergized, the reversing contactor goes to a neutral position where no power is applied to the motor (see Fig. 7.7). Energizing one or the other coil will power the motor in the opening or closing direction. Coil power is usually introduced through a push-button station. To prevent both coils from being energized at the same time, a set of auxiliary contacts is wired in series with the push buttons to open the coil circuit opposite of the one being used. In the neutral position the auxiliary contacts close the circuit up to the push buttons which are normally open. As an additional degree of pro-

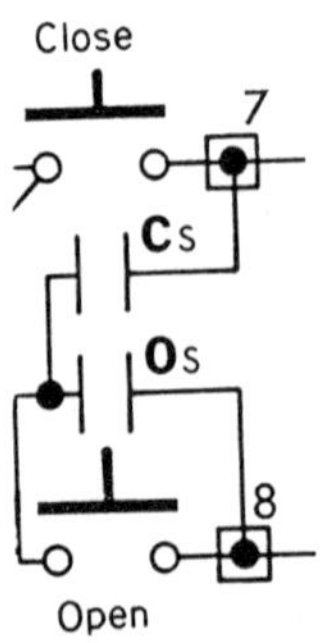

Figure 7.7 Typical holding circuit with momentary contact push buttons. (*Courtesy of EIM Controls.*)

tection each set of contacts is mechanically interlocked to eliminate the chance of both sides powering the motor simultaneously.

Transformers. In a simple single-phase transformer, the iron core is magnetized by a primary winding. When it is fed with an ac voltage, the magnetic flux in the core will be continuously changing as it follows the voltage waveform. A secondary coil wound around the same iron core will also be influenced by this changing magnetic field and will have an alternating voltage generated in it. The relationship between the voltages in the primary secondary windings is directly proportional to the number of turns in each, so it is possible to step up or down the ac voltage to any other level.

Transformers are used with three-phase electric actuators to reduce the motor supply voltage which may be anything up to 575 V and which is much too high for push buttons, limit switches, etc., which require manual access. Lower "control" voltage could be provided as a separate supply but may be economically prohibitive. A much simpler way is to connect a transformer across the motor supply voltage and reduce it to an acceptable level of about 110 to 120 V.

Reversed phase-rotation. When setting the limit or torque switches on a motorized valve, it is possible to run the valve in the opposite direction intended. This causes valves to jam up and motors to overheat because the switches and torque protection are not part of the "live" circuit. The motor will run until it jams the valve with its maximum stalling torque and then overheat until the thermostat trips. This is why the most important start-up check with any motorized valve is the direction of rotation. The best method for doing this is to manually position the actuator so that the valve is in the intermediate position. When power is applied, visually confirm that it is traveling as intended before it is run fully to the open or closed position.

Local-remote control stations. Control stations are used with electric actuators to select local or remote operation, to electrically override the actuator, or when manual operation is preferred (see Fig. 7.8). Local control stations (LCS) are available with switches, push buttons, lights, or locks to meet the exact application requirements.

Switches. Local control stations with switches are usually used with single-phase electric actuators as motor voltage passes directly through them. They are usually *maintained* contacts, which means that once they are manually positioned, they stay there until they are repositioned. For instance, a local-remote switch is a two-position

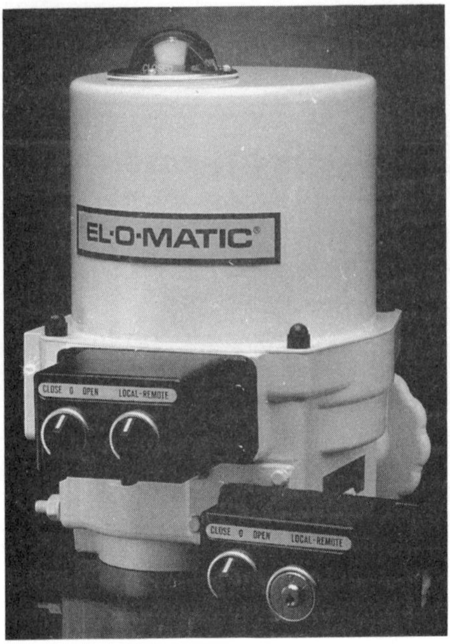

Figure 7.8 Control stations are used to select local or remote operation. (*Courtesy of El-o-matic, Inc.*)

switch that allows local or remote operation. In the *remote* position the actuator control is dictated by the control system or by another control station located remotely. In the *local* position open-close action of the electric actuator is performed at the LCS. In local operation a second switch—usually a three-position switch—provides open-stop-close operation.

Push buttons. Push buttons—and some switches—are available as *momentary* contacts, which means that when pushed or rotated and released, they return to their normal position. Actuator power continues to be applied through a holding circuit that bypasses the push-button contacts. Momentary contacts are usually used with three-phase electric actuators applying control power to a reversing contactor. They are also used as normally closed circuit elements to serve as a stop switch interrupting the holding circuit in the event of an emergency.

Lights. LCS lights may be used to indicate that an actuator is open, closed, or in an intermediate position or to confirm that supply voltage is present.

Locks. Most switches and push buttons are available with padlocks or keys to prevent unwanted tampering.

Motor currents. When sizing control circuits for electric actuators, it is important to consider the operating current they draw individually and collectively. Electric motors draw different amounts of current when they are operating. The *full-load current* occurs when the motor is putting out its full rated power. The *no-load current* occurs when the motor is running with no load attached. The no-load current is usually much less than the full-load current.

When the motor is stalled during operation, the *locked-rotor current* occurs. The name comes from the fact that the motor's rotor is locked or not able to move. The locked-rotor current is even higher than the full-load current. Just how much higher depends upon each particular motor.

Then there is the *starting current.* This is a surge of current that occurs as soon as the motor is switched on. It only lasts for a second or two, or until the motor comes up to speed.

Storage. Electric actuators include geared position limit and torque switches, electric motors, and, in many cases, relays, push buttons, lights, potentiometers, and solid-state circuits. When properly enclosed, they are designed to withstand the effects of driving rain and snow and in many cases to be submersed. However, protection against the effects of condensation created by temperature changes is another story.

Most actuators are supplied with a space heater mounted in the electrical enclosure and, in addition, their enclosure is O-ring- or gasket-sealed. However, prior to installation additional protective steps must be taken at the job site. For short-term storage (less than 1 year from shipment), the following procedures should be followed.

1. *Do not stack.* The actuator should be stored simulating the recommended mounting position; that is, with the motor in the horizontal position and the electrical enclosure either in the horizontal or vertical (up) position.
2. Each actuator should be stored indoors free from excessive dirt, moisture, and temperature changes.
3. If warehousing (as specified above) is not possible, the actuator must be stored off the ground, above possible water or snow level.
4. If item 3 is not possible, the space heater must be energized.
5. If conduit plugs are supplied, do not remove them until electrical hookup is performed.
6. Cover units with a plastic sheet. This serves as protection from rain

and snow. If actuators are stored on gate valves, a stem cover must be installed.

For long-term storage (more than 1 year from shipment) the externally mounted control equipment should be removed and stored in a protected area. The procedures of short-term shortage should also be followed.

7.2 How to Select an Electric Actuator for Multiturn Valves

Manually operated multiturn valves use either handwheels directly engaged with the valve stem or gear operators. In either case the opening and closing action of these operators is unique for each of the two types of multiturn valves—rising and nonrising stem.

Gate and globe valves shut off the flow of a fluid by forcing the closure member (gate disc or globe plug) down into the flow stream. When an electric actuator is used to automate a gate valve, it in effect replaces the valve handwheel to provide the number of turns necessary to completely cycle the valve. It must also provide the torque required to properly act upon the disc through the stem thread thereby torque-seating the valve. For these applications electric multiturn actuators must be designed to withstand the thrust associated with this action. There are many things that complicate the selection of an electric actuator when used with gate valves. First, most gate valves are intended to be manually operated valves and are designed with handwheels that allow operating personnel to apply enough force to fully seat the valve with little effort. However, the way a product is designed is not always how it works in the field. As with all types of valves there are many things that affect its ability to operate smoothly and repeatedly over time. Additionally, there are few if any manufacturers of traditional gate valves who publish operating torques for their manually operated valves. As you will see there are very complicated formulas and necessary safety factors that must be applied when sizing an actuator for the electric actuation of gate or globe valves. The second concern when selecting an actuator is fitting the drive sleeve to the threaded stem of the valve. Especially in the second half of this century many valve manufacturers sourced their products from overseas. In some cases the valves came from more than one source or even many different countries. As a result, for the same size and series valve under a single manufacturer's label the stem dimensions were sometimes different and were therefore not reliable for the manufacturer of mounting hardware. This means that even today some applications of valves of the same size and service may require different size actuators. This is because in order to provide the necessary thrust to properly close and open a valve an electric actuator must use the mechanical advantage of the thread of the stem, which if different, will require different torque values. In some cases the stem diameter may even be too large for the appropriate-size actuator to physically accept.

Sizing guide for electric actuators

Multiturn electric actuator output performance is defined in terms of torque and speed at a specified supply voltage and frequency while subject to thrust reaction when the torque is applied to a threaded stem nut (see Fig. 7.9). For new motorized valves, it is the valve manufacturer's responsibility to specify the operating force required by the valve under the specified service conditions in terms of torque or thrust, or alternatively to confirm acceptance of the electric actuator manufacturer's estimate of performance requirements from given valve and service data.

Screw-operated valves (gate and globe). The size of the actuator required to operate any given value is determined by four parameters.

1. Stem thrust
2. Rising stem diameter
3. Stem torque
4. Operating speed or time

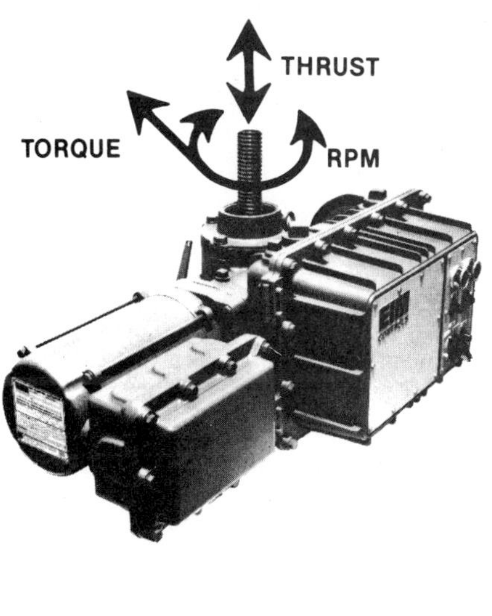

Figure 7.9 Electric actuator for a gate valve produces thrust by applying torque to the threaded stem. (*Courtesy of EIM Controls.*)

The actuator may be sized from the above data, if given, or by a calculation method. The calculation requires more details of the valve design, which can be obtained from the valve manufacturer or by measuring an existing valve.

Large gate valves. Stem thread efficiency must be taken into account when motorizing large screw-operated gates. While a stem of 3-in (76-mm) diameter with a ⅖-in lead may be satisfactory for slow hand operation or short stroke power operation, the continuous energy dissipation from motor operation through a long stroke with heavy gate may cause overheating and rapid wear of stem nuts. Doubling the lead more than halves the power loss for the same travel time, through increased efficiency. Lead approaching one-third the stem diameter is usually best. The total thrust used for actuator sizing must not be less than 3 times the weight of the gate.

Speed specification. Because actuator gear ratios give speed steps, the nearest standard speed to that requested will be offered by multiturn actuator manufacturers. Where it is vital that the travel time not be less than or more than the stipulated figure, this should be stated. There are cases where a combination gear ratio and pulsing power circuit may be used to meet specific speeds.

High operating speeds. Doubling the speed of an actuator for a given torque also doubles the horsepower. However, by doubling the lead of the screw, twice the speed of valve operation can be achieved while retaining the same actuator revolutions per minute and with only a 25 to 30 percent increase in torque and horsepower. To keep the drive self-locking the gate valve threaded stem lead should not exceed one-third of the diameter of the stem.

Self-locking characteristic. In order that the valve remain locked in position when no motion is required, whether in motor or manual drive, the stem thread or gearbox must be irreversible. This requires the consideration of the maximum thread lead which may be used with a given stem diameter. As the lead is increased, the thread efficiency increases to the threshold of dynamic reversibility, where, if motion is started, for instance by a handwheel, the stem thrust from the valve may cause motion to continue. Thread efficiency may also increase to the threshold of static reversibility, where stem thrust from the valve may start stem nut rotation. Some manufacturers use an independent manual drive and do not rely on motor worm gear inefficiency to achieve gate valve stem locking.

Actuator sizing and performance definitions

Seating torque for a gate or globe valve is determined from seating thrust, taking into account the stem thread and coefficient of friction at the stem nut. The former is dimensional; the latter can only be an estimate and is what stem factor (torque or thrust) tables are based upon.

Stem factors

Stem pitch. The distance from a point on a thread to the corresponding point on the next thread (see Fig. 7.10).

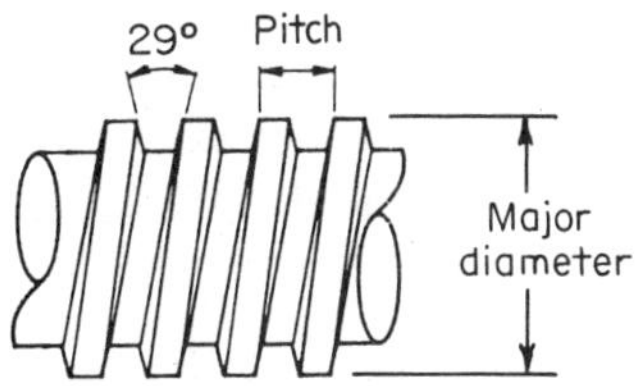

Figure 7.10 Section of threaded gate valve stem. *(Courtesy of EIM Controls.)*

Stem lead. The distance a screw thread advances axially in one turn. The lead and pitch are identical on single-lead, the lead is twice the pitch for double-lead, etc.

Pullout torque [lb · ft (N · m)]. The maximum torque the actuator must deliver during any portion of valve travel. Pullout torque occurs when unseating gate, globe, or sluice gates from the closed position. It is of short duration, typically less than 3 percent of total valve travel.

Run torque [lb · ft (N · m)]. The average torque the actuator must deliver during majority of valve travel. Run torque is generally 10 to 30 percent of pullout torque.

Actuator revolutions per minute. The revolutions per minute the drive sleeve (stem nut) rotates when operating a gate valve. The revolutions per minute determine the time it will take to stroke a valve from full close to open and is a function of the gear train.

Thrust [lb (N)]. The force transmitted to a valve stem by actuator stem nut torque. All multiturn actuators have a maximum thrust rating. The gate valve pullout thrust is the sum of

1. Seat thrust = (valve size)^2 × differential pressure × VF

2. Stuffing box load = stem packing load
3. Stem load = stem area × line pressure

Stem factor. This is the thrust-to-torque conversion factor for 29° Acme stem. It is calculated using stem diameter, pitch, lead, and the coefficient of friction. Stem factor charts are based on a 0.2 coefficient of friction. Lower friction values are possible but must be recommended by the valve manufacturer (0.15 is possible with proper and continuous stem lubrication).

Valve data required. To size a multiturn electric actuator, data should be obtained from the valve manufacturer; however, if the valve is already installed and torque data are not available, calculations must be used to determine thrust and pullout torque. Valve size, stem data, and differential pressure across the valve are required for sizing calculations. Actual thrust calculations may be found in Chap. 6, Sec. 6.3.

$$\text{Seat thrust} = \text{differential pressure} \times \text{seat area} \times \text{VF}$$

$$\text{Pullout torque} = \text{seat thrust} + \text{stuffing box load} + \text{stem load}$$

$$\text{Run thrust} = \text{stuffing box load} + \text{stem load}$$

$$\text{Pullout torque} = \text{total thrust} \times \text{stem factor}$$

$$\text{Run torque} = \text{run thrust} \times \text{stem factor}$$

Always consult the actuator manufacturer for recommendations and stem factor calculations

7.3 Sizing an Electric Rotary Actuator

Electric actuator manufacturers publish output torque values for actuator sizing. Some may publish only one value; some publish two. An electric actuator has basically three important output torque values: start, run, and stall torque. Sizing is an important step in electric actuator selection because undersizing is a common reason for electric actuator failure. Ideally an electric motor should operate under a load using its running torque. This is the motor's natural or nominal output. Luckily enough, most quarter-turn valves and some multiturn valves have torque curves that closely resemble a motor output torque curve. Unfortunately some users of electric actuators are not familiar with motor curves and sizing techniques and may therefore select or specify an actuator that is drastically undersized and consequently has a much lower than expected lifespan.

The electric motor in an electric rotary actuator has an output torque curve that is typical for motors used with this type of device. Its output is calculated by a formula based on power and rotor speed:

$$\text{Torque} = \frac{\text{power}}{\text{revolutions per minute}}$$

This formula indicates that at slower revolutions per minute the motor will generate more torque than it will at higher speeds. For an electric actuator the slowest motor speed occurs when the actuator is at rest and is first energized or when it is in a stall condition. Starting from zero revolutions per minute its speed slowly increases until it reaches its maximum speed. This means that the highest output torque for an electric actuator occurs during the first few degrees of travel and is called the *starting* or *breakaway* torque. At full speed the output drops to what is called the *running* torque. Electric actuator manufacturers publish the starting torque and some include the running torque. The output torque curve for an electric rotary actuator is shown in Fig. 7.11.

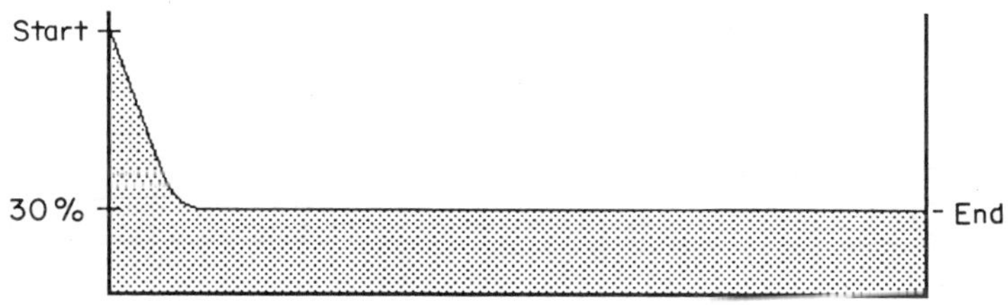

Figure 7.11 Output torque curve for rotary electric actuator.

When sizing an electric actuator, it is important to consider the running torque. For ball and butterfly valves it is appropriate to size the breakaway torque of the actuator with the breakaway or seating torque of the valve. A valve's running torque is similar to that of the electric actuator's output and will therefore properly operate (see Fig. 7.12).

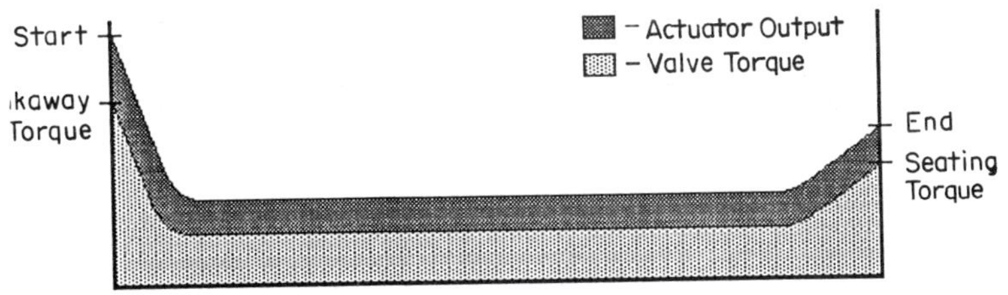

Figure 7.12 Actuator torque curve matched with rotary valve curve.

If an application requires that in the running portion of the output curve an electric actuator must respond to spikes in valve torque, the actuator will increase in output. This is because an increase in the valve torque will slow the motor, and at a reduced speed the output torque will correspondingly increase. Of course, the duty cycle of an electric motor prohibits it from operating continuously beyond its normal running speed and torque, and so sizing for a valve must be done carefully. A plug valve has what is considered a constant-torque requirement value throughout its cycle. To properly size an electric actuator its running torque must exceed the relatively high running torque of the plug valve.

Multiturn actuators for rotary valves

When the operating torque of a rotary valve (ball, plug, or butterfly) exceeds the available output of traditional quarter-turn electric actuators [usually a maximum of about 15,000 lb · in (173 kg · m or 1700 N · m)] a multiturn electric actuator may be adapted to a quarter-turn gearbox (see Fig. 7.13).

As an example some manual quarter-turn gearboxes may be fitted with an electric multiturn drive, after the handwheel has been removed, through a special adaptor flange to serve as the mounting surface. The multiturn input is converted to rotary (quarter-turn) output through the gearing and is controlled by geared limit switches that are part of the electric actuator's control circuit.

The actuator is sized by determining the valve torque and the minimum input torque necessary at the gearbox to cycle the valve. Input speed is usually faster than what is typically used for direct automation of gate or globe valves.

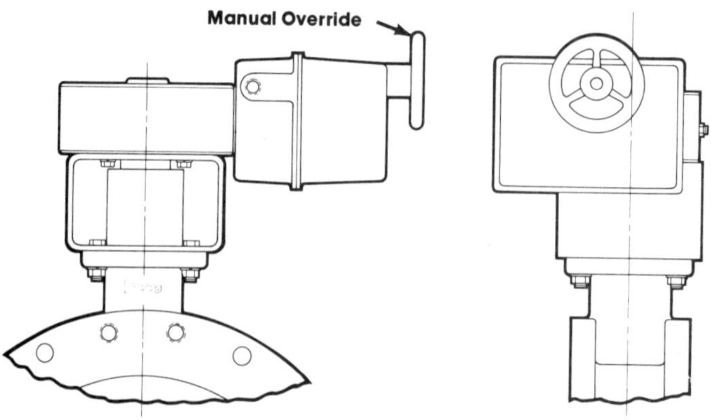

Figure 7.13 Electric actuator with quarter-turn gearbox for higher-torque applications. (*Courtesy of Demco Division of Cooper Industries.*)

Electric actuators are excellent devices for the automation of valves particularly when control system simplicity and product versatility are important. Though it is often tempting to select a smaller, less expensive electric actuator when sizing is concerned, a conservative approach is often the best approach and will result in longer, more reliable performance. Direct computer control, fewer system elements, low maintenance, and energy efficient operation are all features of electric valve automation.

Part

3

Automated Valve Applications

Chapter

8

Actuator Designs for Multiported Valves

In many process systems, multiported valves are used to divert or mix fluids as shown in Fig. 8.1. These valves are available in a number of configurations and flow patterns in three-, four-, and five-way designs. Some of these multiported valves, if automated, require only 90° of rotation to provide the necessary flow pattern. However, the process engineer may find that piping, valving, and related equipment can be eliminated or that a more efficient system can be created if flow arrangements requiring 180° of rotation are installed and automated. The 180° actuators are available to satisfy this requirement and may be either electrically or pneumatically powered.

8.1 180° Electric Actuators

Rotary electric actuators basically are geared motors. Gearing limits the speed and torque but in almost all cases allows continuous rotation. With few exceptions, only the limit switches, each of which is located in series with the motor and supply voltage, are used to cut off the power and limit the rotation. For this reason, many electric actuators are readily adaptable without redesign to 180° operation.

Electric actuators may also be configured to provide three-position rotation (0-90-180°) to control, for instance, the porting of the three-way valve arrangement, which will provide a center position shutoff. This requires extra switches.

8.2 180° Pneumatic Actuators

Pneumatic actuators are also available for 180° rotation. Rack-and-pinion actuators are particularly well suited for this function, as

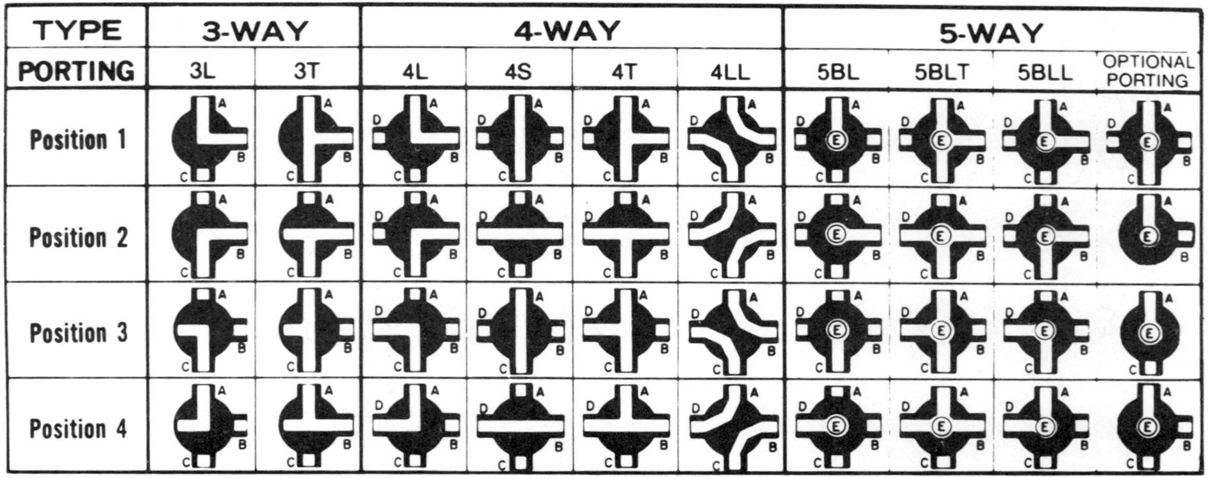

Figure 8.1 Flow patterns for multiported valves. (*Courtesy of Quality Controls, Inc.*)

the relationship between the piston stroke and rack-and-pinion gearing determines the overall rotation of the output shaft. These actuators are also available in both double-acting and spring-return versions.

Three-position pneumatic actuators

Pneumatic rack-and-pinion actuators may also be configured to provide three-position, 180° operation (0-90-180°). This function may be accomplished in at least two ways: 180° actuator with two-stage package and 180° actuator with auxiliary pistons.

Two-stage package. A two-stage package is a control accessory for pneumatic actuators that may be used for 90 or 180° valves. This system is most often used with on-off valves to provide an intermediate stop position between full open and full closed to reduce the flow rate through the valve in filling or batching operations. A two-stage shutoff valve is sometimes called a dribble control package. It is also one of the five basic valve functions mentioned at the beginning of this book.

Two-stage control is accomplished in a variety of ways depending upon specific requirements by using creative pneumatic circuitry. It may include an intermediate stop for open or closed or both and may be used with double-acting or spring-return actuators. The same circuitry—with slight modifications—may also be used with a 180° valve to control three flow positions. In this case the intermediate stop position serves to control rotation at 90°.

Auxiliary piston method. The auxiliary piston design is fully enclosed and is the most compact of all. In design a 180° actuator is fitted with two auxiliary pistons. The piston rod of these assemblies is of such a length as to serve as a mechanical stop for the main power piston at 50 percent of travel. These steps are illustrated below.

The controls required to operate a three-position pneumatic actuator include three 3-way solenoid valves. One for port *A*, one for port *B*, and one for port *C* (port *C* is tapped into each end cap and should be "teed" to pressurize them simultaneously).

Condition I—0 to 90° (Fig. 8.2).

Pressure	Port
Yes	*A*
No	*B*
Yes	*C*

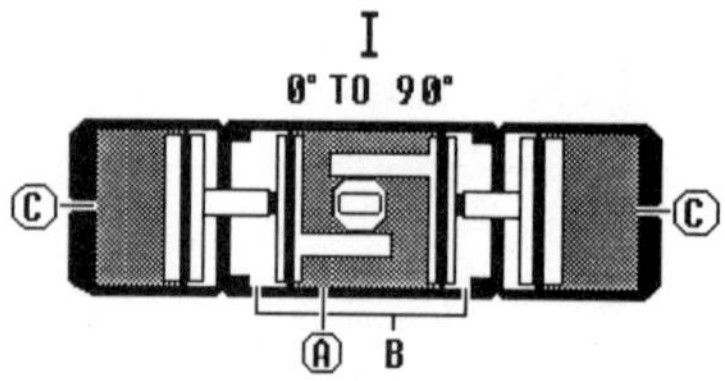

Figure 8.2 Condition I—0 to 90°. (*Courtesy of El-o-matic, Inc.*)

In this condition, the auxiliary pistons are pressurized and stroke inward. The piston rods serve as mechanical stops for the main drive pistons as they stroke outward due to pressure at port *A*. This achieves the 90° position.

Condition II—90 to 180° (Fig. 8.3)

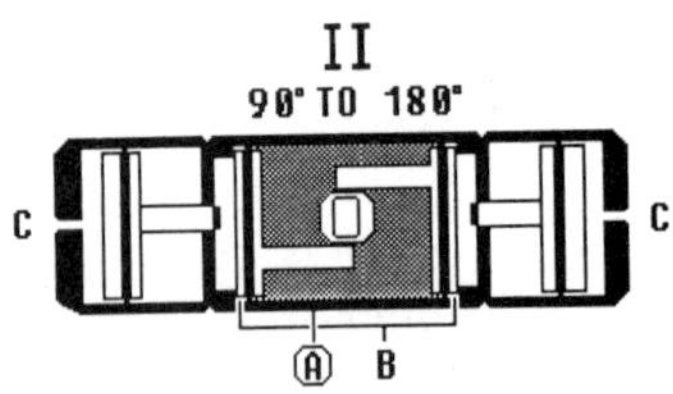

Figure 8.3 Condition II—90 to 180°. (*Courtesy of El-o-matic, Inc.*)

Pressure	Port
Yes	*A*
No	*B*
No	*C*

When port *C* is exhausted, the main drive pistons continue to stroke outward, returning the auxiliary pistons to their original position and at the same time positioning the actuator to the 180° mark.

Condition III—180 to 90° (Fig. 8.4)

Pressure	Port
No	*A*
No	*B*
Yes	*C*

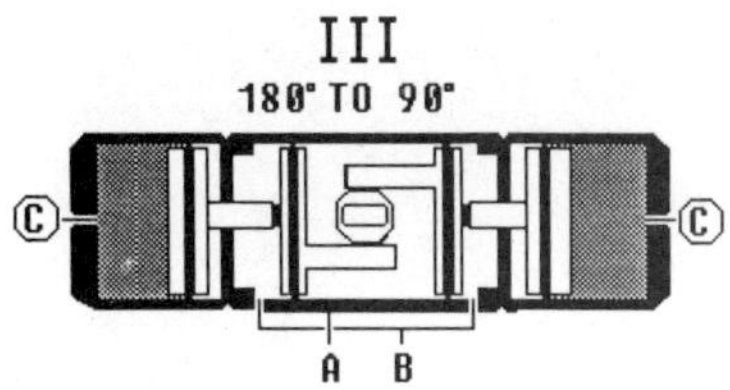

Figure 8.4 Condition III—180 to 90°. (*Courtesy of El-o-matic, Inc.*)

Applying pressure to port *C* only causes the auxiliary pistons to stroke inward and apply force to the main drive pistons. The stroke distance and piston rod lengths are such that they will position the actuator to the 90° mark.

Condition IV—90 to 0° (Fig. 8.5)

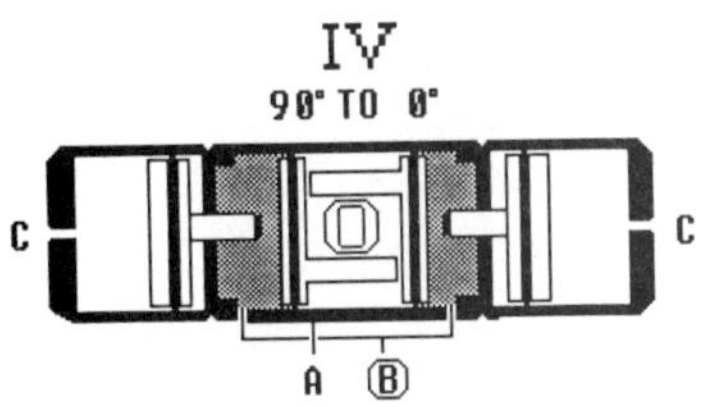

Figure 8.5 Condition IV—90 to 0°. (*Courtesy of El-o-matic, Inc.*)

Pressure	Port
No	*A*
Yes	*B*
No	*C*

To complete the cycle, air is supplied to port *B*. This forces the main drive pistons inward and returns the actuator to the zero degree mark.

Summary. Three-position pneumatic actuators are also available in a fail-safe version to drive a valve to the midposition (90°) in the event of an emergency. At least one unique design uses a push-pull spring arrangement that ensures proper fail-safe positioning whether the valve is in the 0 or 180° position when an emergency occurs.

Chapter

9

Actuator Designs for Fail-Safe Applications

Millions of dollars are spent annually to analyze the safety and reliability of process plants. Expenditures in the chemical industry for risk control have been going up, while property liability insurance premiums have been going down. The trend shows that companies are more willing to spend money to control risk. The payoff is reduced downtime, as well as less likelihood of an accident.

One area of actuation that requires serious review is the emergency shutdown mode. This is a condition that specifies a valve position, in the event of a failure or emergency situation. In most cases, the situation is brought about by a failure of power to the actuator.

9.1 Pneumatic Actuators

For pneumatic actuators, this means that auxiliary power will be required to compensate for the loss of compressed air. This action may be provided in one of two ways: externally, with an air accumulator tank, or internally, with a spring pack.

Accumulator tanks

The accumulator tank method uses a double-acting pneumatic actuator, a variety of sensing controls, check valves, and an accumulator tank (see Fig. 9.1). It is used when a fail position is required from a double-acting actuator that does not have any practical means for spring-return operation. The double-acting actuator for this method must be sized for an air-supply pressure that is less than the available air pressure for the double-acting actuator. This is because stored air pressure is dependent upon volume.

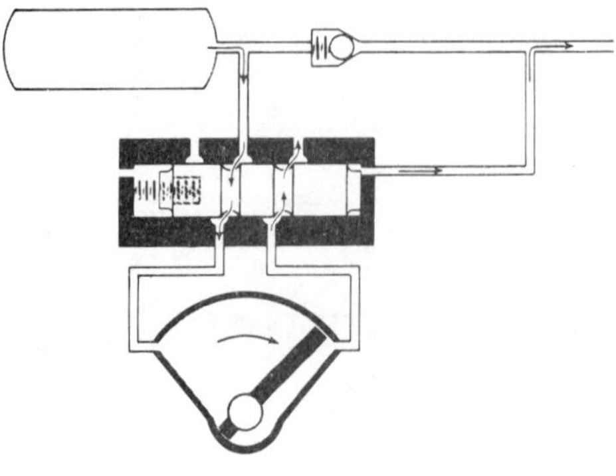

Figure 9.1 Accumulator tank system with vane actuator. (*Courtesy of Xomox Corporation.*)

When air is stored at a constant temperature in an accumulator tank of a volume V_1 at a given pressure P_1 and is allowed to expand into the actuator, which has a volume of its own V_a, to cycle the valve, the new volume for the initially stored air is $V_2 = V_1 + V_a$. Hence, from the formula

$$P_1 \times V_1 = P_2 \times V_2$$

it can be seen that the final pressure P_2 will be less than the initial pressure P_1, so the actuator must be sized for the lower air pressure. Of course, the accumulator tank volume can be increased to reduce the amount of volume change and therefore air pressure loss. These calculations should be performed by the actuator manufacturer. (See Chap. 6 Sec. 6.3 for tank sizing.)

Spring-return actuator designs

The second emergency shutdown method, the internally powered air actuator, requires the addition of a spring-return option. This is accomplished by taking the double-acting actuator and rather than providing air in both directions, having the springs do the work in the fail-safe direction. The actuator is then considered a single-acting, or spring-return, unit.

Spring-return actuators are used in as many as 80 percent of the quarter-turn valve applications in a process plant. When selecting a spring-return actuator, it is important to understand its design and operation to make the best choice. A spring-return actuator consists of an air cylinder and a spring. When air is applied to one side of a piston, it generates a force and at the same time compresses a spring located on the opposite side of the piston. When air pressure is relieved, the spring force pushes the piston in the reverse direction, returning it to its original position.

To further understand the performance of a spring-return actuator, let's review the force output of a simple linear cylinder (see Fig. 9.2).

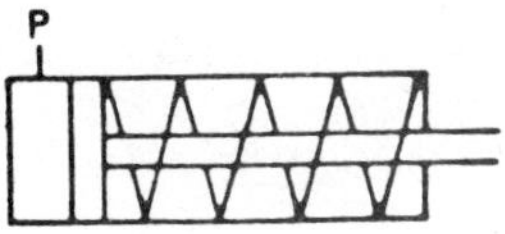

Figure 9.2 Linear stroke spring-return cylinder.

Air stroke. When air is applied to a spring-return cylinder, a force is created. This force is used to do some work (lift, slide, push), but it also compresses the spring and overcomes frictional forces. As the spring is compressed, its resistance to the piston force increases directly with the change in its length. The shorter it gets, the more it resists further compression. This generates a force output curve as shown in Fig. 9.3.

Spring stroke. When air pressure is exhausted, the spring force acts on the piston to generate an output force. In the fully compressed state, the stored energy in the spring is high. As it strokes or is relaxed, the force it generates decreases as its length increases.

Spring designs for rotary pneumatic actuators. In the single-spring design (see Fig. 9.4), the piston is opposed by only one spring, as in our example. To provide the amount of force necessary to do work, the spring selected must be large and of a certain length to provide optimum power throughout rotation.

Spring selection is done by the actuator manufacturer and is based on the spring rate, which is the characteristic of a spring and expresses the change in spring force versus the change in spring length. For a given actuator, the stroke length remains the same, so the spring rate is selected based on the amount of air pressure used and the amount of torque required.

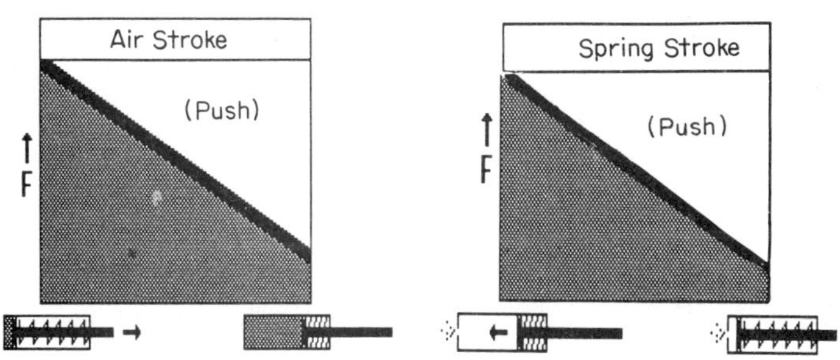

Figure 9.3 Output force curve for linear stroke cylinder.

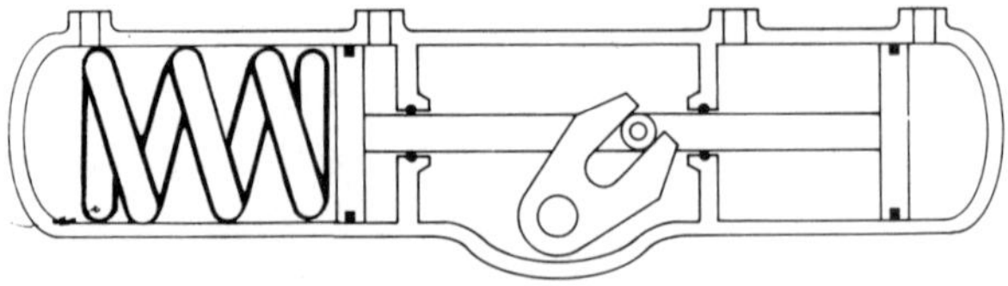

Figure 9.4 Single-spring rotary actuator.

One disadvantage of this design is that the actuator manufacturer may not offer the flexibility (variety of available springs) to accommodate valve torque requirements and air-supply pressures. It is usually necessary to replace the spring with one having the appropriate spring rate for the application. This design also adds significantly to the overall length of the actuator assembly, creating potential installation restrictions and excessive loads on the valve neck and mounting hardware.

Multiple-spring design. The multiple-spring design uses a quantity of springs to generate the required force for an application. Depending on the manufacturer, the quantity can range from 6 to 28 springs. The standard package is usually provided with a maximum number of springs to provide the maximum power for a given model size.

There are two multiple spring methods used for pneumatic actuators: parallel and nested.

Parallel-springs method. The total force generated in a parallel-spring method (see Fig. 9.5) is the sum of the forces of the individual springs. Each spring used in this design is the same size and has the same spring rate for a given model. They are situated next to each other, evenly distributed about the face of the piston. In a typical 14-spring assembly, 7 springs are used on each piston.

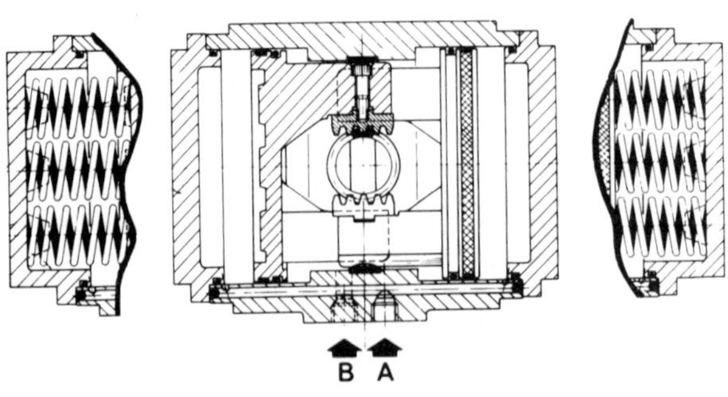

Figure 9.5 Multiple springs are located in parallel on the face of each piston. (*Courtesy of El-o-matic, Inc.*)

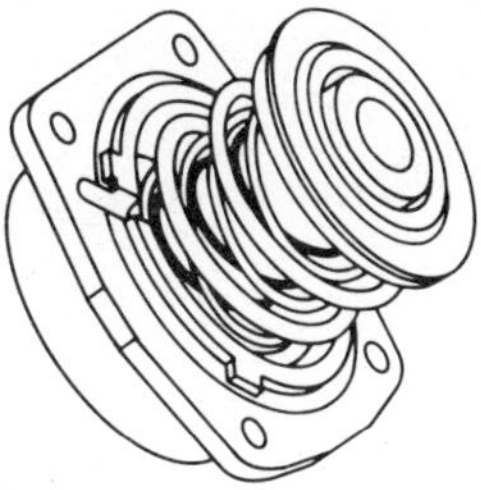

Figure 9.6 Nested springs are fitted within one another. (*Courtesy of El-o-matic, Inc.*)

Nested-springs method. With the nested-springs method, springs are fitted within one another (see Fig. 9.6). The total force generated in this method is also the sum of the individual springs, but it uses fewer springs and each spring has a different spring rate. This adds compactness to the assembly and makes it easier to work with, as the minimal number of springs allows them to be configured in a self-contained spring pack.

In a typical six-spring assembly, three springs are clustered together in a pack and applied to the face of each piston. The most important advantage of the multiple-spring design is that the quantity of springs used may be adjusted to meet specific torque requirements based on available air pressure. It also reduces lead times and adds compactness to the assembly.

9.2 Fail-Safe Electric Actuator Operation

Electric actuators may provide fail-safe action in basically one of two methods: auxiliary power or spring action. The spring-action design consists of an electric rotary actuator that drives a spring into a preload condition. The preload is maintained through an auxiliary circuit that continuously energizes a solenoid and lever assembly. When electrical failure occurs, the solenoid, which itself is spring loaded, releases the lever which in turn releases the main power spring and produces a minimum output of 300 to 600 lb · in (34 to 68 N · m) depending upon the model. This innovative design is very neat and compact but limited in torque.

Some designers prefer a battery backup or auxiliary circuit design for fail-safe action. One method uses a dc motor with a battery backup. A pilot relay senses the loss of electricity and switches over to the battery for supply voltage.

Another design passes dc power through a rectifier circuit to simulate alternating current. This means that ac motors may be used with normally available ac control power even when there is electrical failure.

Electrohydraulic actuators

One of the advantages of electric actuators is that it is not necessary to install and maintain a compressed air system in the field. This means that a totally electric connection from the process control system to the valve is possible. Pneumatic actuators, however, provide a means for fail-safe action which for electric actuators may mean dc motors with battery backup which are both cumbersome and expensive.

With new developments in hydraulic components including electronic-hydraulic devices these actuators are now able to interface with computer-controlled systems for both on-off and modulating applications. Most high-pressure hydraulic actuators are being built on a custom basis, but some hydraulic component advancements have helped make low-pressure systems more practical.

For these applications some manufacturers offer a self-contained electrohydraulic spring-return actuator. As the name implies, this actuator uses an electrical supply to pressurize through a pump the hydraulic fluid required to stroke a spring-return actuator (see Fig. 9.7).

Figure 9.7 Electrohydraulic actuator on quarter-turn valve. (*Courtesy of El-o-matic, Inc.*)

With self-contained electrohydraulic actuators a compressed air system is not necessary and power wiring is like that for a spring-return pneumatic actuator providing, in effect, a fail-safe electric actuator.

Electrohydraulic rotary actuator. A typical electrohydraulic rotary actuator package includes rotary actuator, hydraulic pump, solenoid valve, limit switch, and reservoir. The system operates as follows (see Fig. 9.8).

The electrohydraulic power unit is connected to both sides of the piston chambers of a rotary actuator. When the hydraulic system is energized, pressurized hydraulic fluid enters the spring opposed side of the actuator, causing the two pistons to start transmitting a rotary motion to the central drive shaft by means of a rack and pinion. As the actuator moves through 90° the hydraulic oil on the low-pressure side of the pistons (spring side) returns to the power unit reservoir maintaining a constant level of oil and providing a totally sealed hydraulic system. When the actuator is in the desired position, the hydraulic pump within the power unit is deenergized through limit switches, and the actuator is held in position by locking pressure within the unit.

The return direction is achieved by deenergizing the normally open solenoid vent valve which allows oil in the pressure chamber to flow back into the power unit reservoir, thus reducing the pressure and allowing the opposing return springs to force the pistons together and provide a reverse rotary motion.

Electrohydraulic actuators when fitted with the appropriate controls (positioner and feedback) are ideal for modulating applications

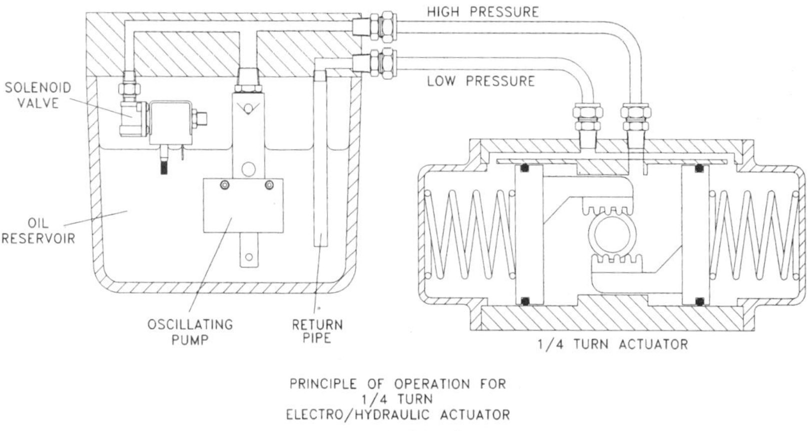

Figure 9.8 Operating schematic of electrohydraulic actuator.

because the hydraulic fluid offers infinite stiffness for stability. Electrohydraulic actuators are available for linear or rotary output. In modulating control applications an electronic interface unit (positioner) will accept an analog signal representing the desired valve position and a signal from the valve position feedback potentiometer representing the actual valve position. The unit will detect an out-of-balance condition and will switch the hydraulic pressure until the system is at balance.

Specifications for an electrohydraulic power unit

Enclosure: Cast aluminum, or iron or steel weatherproof. Designed to meet NEMA IV or VII

Operating temperature: −4 to 131°F (−20 to +55°C)

Mounting: Vertical or horizontal

Cable entry: Two conduit entries threaded ½ in NPT (M20)

Electrical supply: 240 or 110 V ac + 10 to 15 percent 50/60 Hz

Input signal: 4 to 20 mA, 0 to 20 mA, or 0 to 1 V dc (for control)

Power consumption: Pumping 275 V · A, reducing to 7 V · A when solenoid bleed valve only energized

Accuracy (overall): <1 percent

Linearity: <0.75 percent

Hysteresis: <0.1 percent

Repeatability: <0.4 percent

Linear electrohydraulic actuator

Thrust: 150 to 1500 lb (667 to 6672 N)

Stroke: Up to 3½ in (88.9 cm)

Action: Direct-acting (release pressure; actuator stem retracts); reverse-acting (release pressure; actuator stem extends)

Rotary electrohydraulic actuator

Torque: 185 to 10,600 lbf · in (20.9 to 1198 N · m)

Angle of travel: 0 to 92°. Adjustable between 80 to 92°

Spring return: Clockwise or counterclockwise

9.3 When to Use Fail-Safe Actuators

There are countless applications where a fail-safe actuator should be used. The following table gives examples of such applications.

Fail open	Fail closed
To prevent pressure buildup of a system or component due to a power failure to the valve actuator	To prevent pressure buildup of a system or component due to a power failure to the valve actuator
To prevent pressure buildup of a system or component due to a power loss of a system pressure, using a pilot-operated valve or actuator of that system	To close a valve because of a leak or rupture in the line
To allow toxic gases or liquids to vent in case of power failure	To prevent fueling of a fire, such as in oil companies or chemical plants, when actuator power failure is possible because of that fire
To allow coolant to circulate in a system or component, such as in the heat exchanger of a nuclear power reactor, or in the event of cycling or power failure of the actuator power source	To prevent toxic gases or liquids from escaping because of a system power failure

It should be noted that fail-safe actuators may also be used in conjunction with other systems when equipped with signal or sensing devices for either fail-open or fail-closed applications. The following is a list of some of these.

1. High or low liquid level relief
2. Runaway pump or motors
3. Overheating or cooling
4. Low or high pressure relief
5. Dangerous liquid or gas handling
6. Life support with liquids or gases, such as water or oxygen

Another shutdown condition that may be considered in the event of an emergency is the *as-is* mode. This calls for a pneumatic actuator or an electric actuator to remain in place upon power failure.

For some valves, which have a high natural torque, this may be accomplished with a simple double-acting actuator and no further pneumatic controls. These valves are not usually subject to movement under process flow conditions. Ball and plug valves are considered to have high natural torque. A butterfly valve, however, may cycle under high-velocity flow conditions when air pressure to the actuator is lost.

In this case a pneumatic lockup system is required. The lockup system consists of a spool-type valve that allows air to operate the actuator through open porting in normal operation. It has a sensing diaphragm that is spring-loaded. When air pressure drops or a failure is detected, the spool shifts and captures the air pressure in both chambers of the actuator to prevent operation.

Electric actuators use gear designs that may or may not be inherently self-locking. Spur gears usually must be fitted with an electromechanical brake for as-is failure positioning. Most other designs are self-locking.

The pneumatic actuator examples above dealt specifically with an air failure condition. Some conditions, however, may require that an emergency shutdown occur upon electrical failure. In this case, a single-acting solenoid-operated pilot valve, interfaced with a spring-return actuator or a double-acting actuator if air pressure is guaranteed, usually satisfies this condition.

Actuators used for an emergency shutdown may also be configured to control, signal, or pilot other devices and components in the process. By adding a rotary switch to a pneumatic or electric actuator, it is possible to sound an alarm or shut down a pump, fan, or other motor when the actuator goes to the failure position.

Directly related to the emergency shutdown mode methods and requirements is the ability to operate a valve in the event of actuator or power failure. When it is necessary to guarantee the operation of a valve in the event of these failures, it is important to analyze the various methods available to manually override an actuator in the field. These methods are covered in Chap. 6.

Chapter

10

Actuators for High-Temperature or Fire-Safe Applications

A carefully selected fire-safe valve may also be considered fail-safe. In this case *fail-safe* refers to the failure or disintegration of a soft seat when subject to a fire and the subsequent safe condition that develops when some second-stage sealing condition occurs. Examples of this are metal-to-metal seating against the machined body of the valve or a secondary metal ring built into a butterfly valve seat.

Of course, if a valve is not already in the desired safe position (open or closed) when a fire occurs, provisions must be made to ensure that it can be powered to that position in case of an emergency. High-temperature or fire-safe actuators and coating systems, and heat-sensing accessories, are designed to provide for this action.

10.1 Spring-Return Lever and Link for Manual Valves

A spring-return lever is an operator that allows a rotary valve to be opened or closed manually but that returns the valve to a fail-safe position once released. Because of the amount of torque required to both operate a valve and compress the spring, spring-return levers are usually limited to use with 2-in (50-mm) valves and smaller.

For fire-safe applications the spring-return lever may be fitted with a fusible link. A fusible link is made of a blend of zinc and lead or other metals with a relatively low melting point. The metals are blended in proportion to the desired melting point but are strong enough at ambient temperature to hold the spring-return level in a preloaded position for normal process conditions. When the melting

Figure 10.1 Spring-return handle with fusible link. (*Courtesy of Pittsburgh Brass Mfg. Company.*)

point is reached during a fire, the link melts and releases the lever. Spring action then drives the valve to the predetermined fail-safe position (see Fig. 10.1).

10.2 Thermal-Sensing Manual Gear Operator

For larger valves requiring higher operating torque a thermal-sensing gear operator may be used. This gear operator uses a spring-wound gear segment which is activated when fusible links melt after exposure to high temperatures. The spring-return action can be activated by remote electric, pneumatic, or hydraulic signals or may be operated manually. After exposure to fire or remote activation, the totally enclosed gear segment is disengaged to allow the spring to return the valve to its fail-safe position, either open or closed. During normal operation the unit performs as a quarter-turn manual gear actuator which can be set at any position, from 0 to 90°, for throttling and on-off control. It does not require an initial setting of the failure mechanism. The actuator gear is automatically declutched when activated and is completely enclosed to eliminate handles that might injure employees. The housing and cover of the actuator are ductile iron, and the gear segments are stainless steel. The spring is made of heat-treated alloy steel.

10.3 Fusible Elements for Pneumatic Actuators

Pneumatic spring-return actuators are available for linear or rotary valve automation. By design they power a valve to a desired fail-safe position upon loss of supply air pressure. In high-temperature or fire-safe applications standard construction spring-return actuators may be used to provide one spring stroke if the following are considered:

1. Maximum operating temperature for one stroke
2. Maximum operating temperature of actuator
3. Duration of the presence of maximum temperature before spring-return action must occur

If it is determined that a standard pneumatic actuator is adequate to meet these requirements, the actuator may be fitted with a fusible plug to respond to a fire. The fusible plug, like the fusible link, is a zinc and lead blend that melts at a specific temperature. A tee fitting is installed on the actuator pressure inlet port and the branch of the tee is fitted with a fusible plug. When exposed to a fire the plug melts and represents a break in the air-supply line which exhausts air pressure from the actuator and provides for the necessary spring action (see Fig. 10.2). If the temperature at which the air-supply line break is to occur is not critical, some users elect to install the air system using polypropylene tubing which melts when exposed to fire.

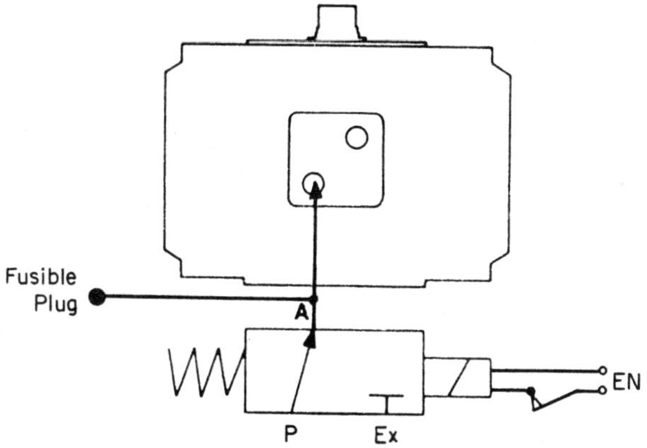

Figure 10.2 Fusible plug is used to exhaust actuator pressure when high temperature is sensed.

For simple, fail-safe action, each of the previously discussed systems can meet some or all of the requirements for some fire-safe systems. There are situations, however, where it may be necessary to cycle a valve during or after exposure to a catastrophic fire condition. Actuator manufacturers offer two systems for use here. One is a pneumatic actuator constructed of materials and seals that can withstand a fire, and the other is a protection system that prevents the actuator from reaching a temperature where seal failure or component distortion may occur.

10.4 Fire-Safe Actuators for Quarter-Turn Valves

Unlike fire-safe valves some actuators may not survive a fire, but if they can provide the necessary spring action before temperatures become damaging, this may prove to be adequate for most applications.

Manufacturers of fire-safe valves and/or actuators are careful to remind users that the term *fire safe* is difficult to define and has no truly universal interpretations. What may be considered an extreme condition for one user may be the norm for another.

Actuator manufacturers consider the degree to which a fire can affect the performance of an actuator during and after a fire. The expected term of operability after the fire and the temperature and duration of the fire are critical to the selection of an actuator.

The term *fire safe* has become a generic term used in the marketplace with regard to any fire cycle life in a valve and actuator which would function during exposure to a fire situation. To assist in clarifying the crossover of numerous definitions, the basic definitions around the performance standards of a fire-safe actuator system are as follows.

1. *Fire exposure:* Fire exposure is a condition where a valve and actuator have been directly subjected to an open fire of sustained time and temperature. The actuator will demonstrate a degree of operability and tightness as described by the following terms.

2. *Fire-resistant actuator:* The actuator shall operate satisfactorily under ambient temperature conditions of 300°F (150°C) and lower prior to exposure to a fire. When exposed to fire, the actuator will resist the effects of high-temperature exposure for a specified period of time but may not remain operationally functional during or after the event. The actuator will probably be permanently damaged and require repair or replacement.

3. *Fire-safe actuator:* The actuator shall operate satisfactorily under ambient temperature conditions of 300°F (150°C) and lower prior

to exposure to a fire. When exposed to fire, the actuator will remain functional for multiple cycles during the event. When the fire has been extinguished, the accumulative damage and deformation caused by high-temperature operation and thermal shock shall be such that the actuator will remain fully operable for at least one cycle. Major repair or replacement of the actuator may be required.

4. *Fireproof actuator:* The actuator shall operate satisfactorily under ambient temperature conditions of 300°F (150°C) and lower prior to exposure to a fire. When exposed to fire, the actuator will remain functional for multiple cycles during and after the event. Fire exposure will result in minor or no internal damage to the actuator.

Fire-safe pneumatic actuators are designed for a standard operating temperature up to 300°F (150°C); however, if exposed to a fire of 2000°F (1110°C) for 15 min, it should remain functional during and after the fire. Typically the criterion is that the actuator operate for at least three full cycles when exposed to a fire. It must also remain functional while subjected to a water spray at 5 and 20 min after ignition and for at least one cycle after the fire is extinguished.

These actuators are constructed of austenitic stainless steel for corrosion protection and to exhibit high strength at elevated temperatures (see Fig. 10.3). Piston and shaft seals are usually made of a resilient material (Buna or Viton) for normal operation plus a second

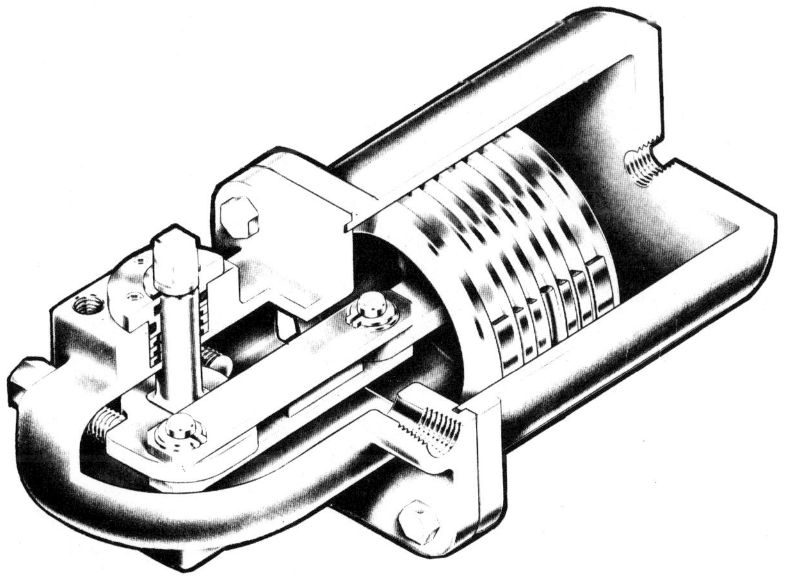

Figure 10.3 Fire-safe pneumatic actuators are designed to remain operational after exposure to 2000°F fire. (*Courtesy of Randon Research.*)

sealing system that utilized a composite, labyrinth seal for operating in temperatures up to 2000°F (1100°C). To prevent seal and mechanical seizure all moving parts are lubricated by a blend of heat-resistant synthetics. Fire-safe pneumatic actuators are available in double-acting and spring-return versions up to 100,000 lb · in (11,300 N · m).

10.5 Fire-Safe Enclosures and Coating Systems

When an actuator required for fire-safe applications is unavailable, too expensive, or for whatever reason not practical, some users will specify a fire box or coating system. Either of these methods is intended to give the actuator (or instruments) thermal protection. When external temperatures reach fire-safe levels, the actuators are subjected to dramatically reduced temperatures that are within their maximum tolerances because of the thermal properties of the fire box.

In some situations a fire box may be as much as 10 times the size of the actuator it is protecting. Stainless-steel construction and elaborate installation may also make such a method less appealing. In these cases one may select a coating system for fireproofing.

10.6 Fireproof Coatings for Valve Actuators

A fireproof coating for a valve actuator limits the temperature rise inside the actuator to not more than 140°F (60°C) for at least 15 min when directly exposed to a 2000°F (1100°C) "high rise" petroleum fire with a contained heat flux of at least 55,000 BTU/(ft^2 · h).

How it works

When the normally passive coating is exposed to fire, the surface of the coating intumesces (swells) forming an insulating char under a glazed surface. As the heat penetrates this insulating char, more passive material becomes active and intumesces in another expansion, forming more and deeper char. The surface reaches flame temperature and reflects heat back into the fire, while the chemical reaction of the intumescing, subsurface material is cooling the metal beneath. The cycle continues until the material is fully intumesced or the flame is extinguished.

Fireproof coatings may be used on pneumatic or electric actuators but must be applied through a molding process and so do not lend themselves to field installations. When selecting a valve-actuator package for fire-safe applications, one should also consider the mounting hardware's ability to withstand high temperatures and thermal shock to ensure operation. Finally, the drive coupling should be designed to accommodate the upward motion of the valve stem when line pressure forces it up against the metal body once the stem thrust washer has disintegrated. This motion may be as little as ⅛ in (3 mm) but is necessary to prevent stem area leakage.

Chapter

11

Hazardous-Area Classification and Equipment

The term *hazardous location* once meant a place within a petrochemical or some other equally volatile plant. Today, hazardous manufacturing processes include coatings, adhesives, and flammable liquids in parts-cleaning tanks and dry-cleaning plants. Even food processing and other agricultural-related industries are becoming familiar with *hazardous* designations by the nature of the processes involved, such as grain elevators or flour and feed mills. Textile mills and any plant that creates sawdust also fall into the hazardous category.

Selecting equipment that may be used in such locations requires an understanding of how hazards are classified. The NEMA, Underwriters' Laboratories, and the National Fire Protection Association have categorized hazardous environments by the following classes with divisions similar to European *zones* (see Table 11.1).

The *division* defines the probability of an explosive mixture being present:

Division 1 locations are areas where the hazardous concentration exists continuously, intermittently, or periodically under normal operating conditions.

Division 2 locations are those where the hazardous vapors are present only in cases of accidental rupture or breakdown of equipment.

Classification of hazardous areas varies slightly between the United States and other countries. The European classification of hazardous areas generally follows that of the International Electrotechnical

TABLE 11.1 Hazardous Locations

Class I	Locations in which flammable gases or vapors are or may be present in the air in quantities sufficient to produce explosive or ignitable mixtures
Class II	Locations which are hazardous because of the presence of combustible dust
Class III	Locations which are hazardous because of the presence of easily ignitable fibers or flyings but in which such fibers or flyings are not likely to be suspended in air in quantities sufficient to produce ignitable mixtures

Commission (IEC) and is affiliated with the International Organization for Standardization (ISO), as its electrotechnical division. The IEC classifications are as follows. A hazardous area is divided into zones:

Zone 0: where an explosive gas-air mixture is continuously present for long periods

Zone 1: where an explosive gas-air mixture is likely to occur in normal operation

Zone 2: where an explosive gas-air mixture is not likely to occur, and if it occurs, it will only exist for a short time

The following table shows the zone and division cross reference used in IEC countries and North America.

<table>
<tr><th>IEC</th><th>Zone 0</th><th>Zone 1</th><th>Zone 2</th></tr>
<tr><td>United States</td><td colspan="2">Division 1</td><td>Division 2</td></tr>
</table>

The responsibility for the area classification of a plant rests jointly with the engineering, process, safety, and any other departments involved in its design and operation. Factors to be considered include

1. The probability that hazardous gas is present
2. The quantity of hazardous gas
3. The nature of the gas (e.g., is it heavier than air?)
4. The extent of ventilation
5. The consequences of an explosion

11.1 Product Classification

Two other safety aspects are commonly included in the classification of devices for use in hazardous areas—gas groupings and temperature classification. The gas grouping takes account of the maximum amount of energy which can be released under operating or fault conditions, whereas the temperature classification is concerned with the maximum temperature which can be attained by the external surface of the device.

Temperature classification

Gas-air mixtures can be ignited by contact with a hot surface, and consequently all electrical equipment used in hazardous atmospheres must be classified according to its maximum surface temperature. Table 11.2 shows the ignition temperature to be taken into consideration in Europe (IEC) and the United States (NEC) for the gases and vapors stipulated there. All temperature classifications, unless otherwise specified, are assessed with reference to a maximum ambient temperature of 40°C (104°F). If the equipment is used in a temperature higher than this, then its temperature classification should be reassessed.

TABLE 11.2 Temperature Classes

IEC		NEC			Ignition temperature of gases or vapors, °C
Temperature class	Max. surface temperature, °C	Temperature identification number	Max. surface temperature °C	Max. surface temperature °F	
T1	450	T1	450	842	>450
T2	300	T2	300	573	>300
		T2A	280	536	>280
		T2B	260	500	>260
		T2C	230	446	>230
		T2D	215	419	>215
T3	200	T3	200	392	>200
		T3A	180	356	>180
		T3B	165	329	>165
		T3C	160	320	>160
T4	135	T4	135	275	>135
		T4A	120	248	>120
T5	100	T5	100	212	>100
T6	85	T6	85	185	> 85

TABLE 11.3 Gas Groupings for IEC*

Example of gases	IEC	North America (group)
Hydrocarbons such as alkenes, including propane, benzenoids, alkenes, gasoline *Oxygen compounds* such as carbon monoxide, alcohols and phenols, some aldehydes, ketones, esters *Halogens* *Nitrogen compounds* such as ammonia, amines, amides	IIA	D
Natural gas *Hydrocarbons* such as ethylene propylene *Oxygen compounds* such as ethyl ether, aldehydes *Hydrogen sulfide*	IIB	C
Hydrogen Carbon disulfide	IIC	B
Acetylene (with special material limitations)	IIC	A

*NOTE: These are approximate correlations.

Gas or apparatus grouping

The same gas groupings are used for flameproof and intrinsically safe equipment, and tables are available showing the equipment classification which can be used with particular groups of gases. Table 11.3 gives some examples with the IEC and North American classifications. Gas groups E, F, and G are concerned with hazards associated with the presence of dust.

The overall term used in Europe covering all methods of protection is *explosion-proof,* and the symbol Ex is used. In the United States engineers tend to use the term explosion-proof as being synonymous with the European term *flameproof* and this often causes confusion.

11.2 Methods for Safe Control

Understanding design methods for operating equipment in hazardous areas starts with the combustion triangle (see Fig. 11.1). Fuel, oxygen, and a source of ignition (spark or temperature) must be present at the same time (and in the necessary proportions) for combustion to occur.

One approach to creating a safe environment for hazardous areas is *confinement*; isolating the area reduces the possibility of accidents. Isolation involves removing or confining any possible element which could create a spark and ignite an explosion. There are three common

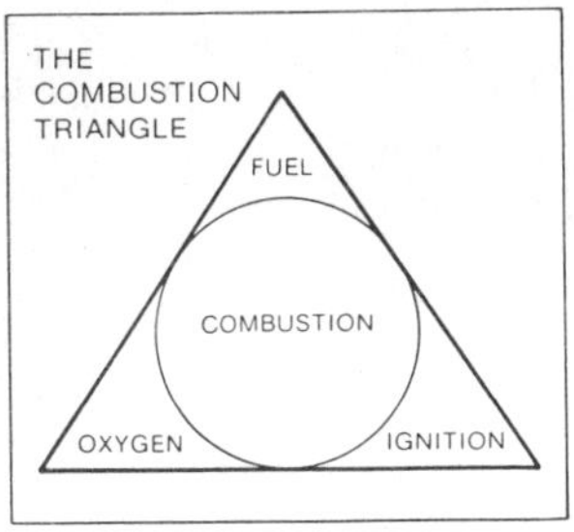

Figure 11.1 The combustion triangle.

methods of providing safety within a hazardous location, categorized by the power technology used: (1) pneumatic, (2) explosion-proof, and (3) intrinsically safe systems.

Pneumatic systems

Pneumatic systems are, by their nature, a safe means of control because they are powered by air. Pneumatic systems are generally clean and easy to service, but the number of control operations performed by pneumatic sequencers is limited. This can be augmented by an electropneumatic interface which enables pneumatics to be controlled by a Programmable Logic Controller.

The drawbacks to pneumatic systems are in distance and reaction times. Where installations are spread over a wide area, the slow reaction time of pneumatic systems increases control reaction time. The length of control circuits in a total pneumatic system must be limited for the control cycle time to remain short.

Explosion-proof housings

Explosion-proof housings provide a simple means to adapt electric and electromechanical and electropneumatic controls to hazardous locations. Explosion-proof housings are designed to withstand the explosion of a mixture inside the enclosure and to prevent the spread of the flame to the outside. These enclosures are effective, especially for interrupting high currents to motors using limit switches. However, this method lacks flexibility in the use of sensing techniques because of the size of the devices. In addition to the space required for explosion-proof devices, material and labor costs for installation and service may be high.

NEMA classifications. The NEMA classifies electrical equipment and enclosures for industrial applications as follows:

I. General purpose	Protects against indirect splashing of dust and light but is not dusttight; primarily prevents contact with live parts; used indoors and under normal atmospheric conditions.
II. Driptight	Similar to type I, with addition of drip shields or equivalent; used where condensation may be severe as in cooling rooms and laundries.
III. Weather resistant	Protects against weather hazards such as rain and sleet; used outdoors on ship docks, for construction work, and in tunnels and subways.
IV. Watertight (weatherproof)	Must exclude at least 65 gal/min (247 L/m) of water from 1-in (25-mm) nozzle delivered from a distance of not less than 10 ft (3 m) for 5 min. Used outdoors on ship docks, in dairies, and in breweries.
V. Dusttight	Provided with gaskets or equivalent to exclude dust; used in steel mills and cement plants.
V. Submersible	Design depends on specified conditions of pressure and time; used for submersion in water, as in quarries, mines, and manholes.
VII. Hazardous location (explosive gas or vapor)	Meets application requirements class I of National Electrical Code; conforms with specifications of Underwriters' Laboratories, Inc.; used for atmosphere containing gasoline, hexane, naphtha, benzine, butane, propane, acetone, benzol, lacquer-solvent vapors, and natural gas.
Type 8 enclosure (hazardous locations, oil-immersed)	Type 8 enclosures are for indoor or outdoor use in locations classified as class I, Groups A, B, C, or D as defined in the National Electrical Code.
Type 9 enclosure (hazardous locations, dust-ignition proof)	Type 9 enclosures are for use in indoor locations classified as class II, groups E, F, or G, as defined in the National Electrical Code.
IX. Hazardous locations (combustible dust)	Meets application requirements class II of National Electrical Code; conforms with specifications of Underwriters' Laboratories, Inc.; used for atmospheres containing metal dusts, carbon black, coal or coke dust, flour, starch or grain dusts.

Intrinsic safety

Intrinsic safety implies that there is insufficient electric energy in a circuit to ignite the most readily flammable mixture of a gas and air. As mentioned in relation to classification, the energy can appear as a spark resulting from its sudden release or in thermal form from the temperature rise of a surface. The equipment design must consider fault conditions as well as normal operation.

According to the National Electrical Code, "Intrinsically safe equipment and wiring shall not be capable of releasing sufficient electrical or thermal energy under normal or abnormal conditions to cause ignition of a specific flammable or combustible atmosphere mixture in its most easily ignitable concentration."

There is a fundamental difference between intrinsic safety and the other equipment application techniques. None of the others aims at preventing release of energy at a dangerous level. Instead they avoid the possible effects in a number of ways:

1. Ensuring that when a spark occurs, it is prevented from reaching an explosive mixture
2. Ensuring that any explosion is contained
3. Reducing the hazard by diluting the gas mixture to a safe level
4. Protecting against excessive temperature or spark which occurs exceptionally

Intrinsic safety, on the other hand, deals with the root cause and ensures that there is insufficient energy available, whatever happens, to cause an explosion. It is, therefore, considered by some to be the safest and most technically elegant approach.

It has a number of practical advantages, offering compact design; reliability; low cost; simple installation; and the possibility, not available with flameproof equipment, of on-line maintenance, if required. The principal disadvantage is the relatively low amount of power available, although this has been largely overcome by a growing range of specially designed components and equipment. For all practical purposes, intrinsic safety is the only safe technique suitable for zone 0 and is often the preferred approach for zone 1 (division 1).

There are two standards on intrinsic safety in IEC countries, "ia" being the higher standard where safety is maintained with up to two faults and "ib" where safety is maintained with up to one fault. Equipment certified to ib standards is generally acceptable in all zones except zone 0, and ia equipment is suitable for use in all zones. In America there is only one standard, and all hazardous area equipment must maintain safety with up to two component faults.

Intrinsically safe elements in an automated valve package

Background. Pneumatic actuators for process valve automation are well suited for low-power or intrinsically safe system integration. Where electric motors or other high-power devices do not directly interface with today's computer-controlled systems, pneumatic actuators when properly sized and fitted with compatible control accessories can satisfy a variety of computer-direct process control requirements. When used in an intrinsically safe system, pneumatic actuator control accessories must be connected to a properly selected and approved barrier.

What are intrinsic safety barriers? Intrinsic safety barriers are current- and voltage-limiting assemblies which are designed and constructed to requirements as described by ANSI/UL 913 (formerly NFPA 493) in the United States and CSA 22.2 number 157M in Canada. When operating conditions are abnormal, the intrinsic safety barriers circuit is designed to divert excessive current to ground (see Fig. 11.2). This diversion is accomplished by the Zener diodes, while current limiting is provided by a resistor. If voltage higher than the maximum allowed

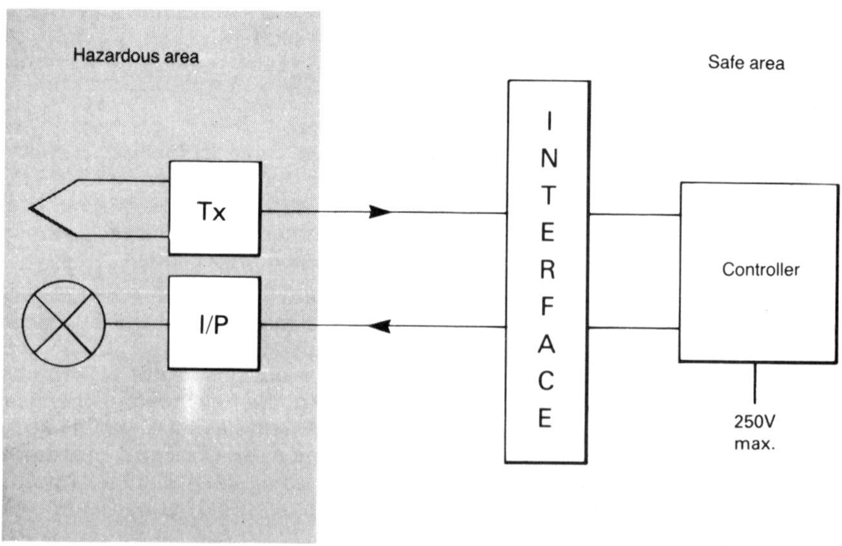

Figure 11.2 Typical intrinsic safety barrier interface circuit.

by the safety barrier is applied across the barrier, the Zener diode would then conduct that excess current to ground. Should current continue to increase, the fuse will open the circuit arresting current flow altogether.

Advantages of low power. All other methods of protection—e.g., pressurization, use of explosion-proof enclosures, or oil filling—rely on the maintenance of a physical barrier between the explosive atmosphere and the electric circuit. Out on the plant and difficult to monitor, a hazardous area enclosure has only to be breached at one point for protection to become nonexistent. In contrast, intrinsically safe elements provide inherent protection by restricting the energy at its source and as a result offer some advantages that include

1. *Economy.* Enclosures are lighter, less cumbersome, and economical. Costly screwed conduits can be replaced by ordinary wiring. Thermocouples, resistors, switches, and other nonenergy-storing field equipment can be constructed to ordinary (weatherproof) specifications.
2. *Live maintenance.* It is not necessary to cut power before calibrating or otherwise adjusting field equipment. Glanded or magnetically coupled controls through flameproof housings are not needed.
3. *Reliability.* The system remains safe if seals fail, cables are severed, or the covers of enclosures or conduit boxes are improperly replaced. Switches do not require long, thin flame-retarding air gaps which are prone to corrosion and seizure.
4. *Safety.* Personnel cannot be harmed by the low voltages used in intrinsically safe circuits.

There are three common electric components that are used with pneumatic valve actuators: the solenoid valve, limit switch, and transducer (or electropneumatic positioner). Each of these devices is manufactured in a version that allows them to be classified as low-power and are therefore appropriate elements for use in an intrinsically safe system.

Certification in the past was applied to intrinsically safe systems as a whole, including the safe-area equipment. This was inflexible and restrictive, eliminating choice and making it difficult to change any item. But now all certifying authorities have adopted a modular approach, known in the United States as the *entity concept,* which allows the user to construct safe systems in-plant, using equipment from any manufacturer, provided that it is electrically compatible and certified to the same standard.

1. Solenoid-operated pilot valves for intrinsically safe systems

General description. Intrinsically safe pilot valves are ideal for applications where the hazards of combustible atmospheres exist or where low-power circuits are a desired feature. When properly installed with approved intrinsic safety barriers, the risk of explosion due to electric or thermal energy is virtually eliminated. The intrinsic safety design is achieved through an operator design requiring minimal current for operation.

Intrinsically safe operator. An intrinsically safe operator is designed to keep current draw to a minimum (29 mA) by accurately controlling plunger strokes. Also, since the pilot orifice is located in the sleeve rather than in the valve body, removal of the operator from the valve does not alter stroke tolerances or current draw. This operator design permits the valve to be operated in any position. This is particularly important for a pneumatic actuator pilot valve when the installed orientation of the process valve cannot always be assured.

Electrical data

- Coil, circuit, and other parts are epoxy encapsulated for system integrity.
- Housings are made of either polyamide and fiberglass or epoxy-coated metal. All housings are waterproof construction meeting NEMA 4 requirements.
- Coil resistance is 298 Ω at 20 mA.
- Maximum voltage is 30 V dc.
- Minimum operating current is 29 mA.
- Unprotected capacitance is 0.
- Temperature rise is less than 37°F (3°C).
- Maximum ambient temperature is 149°F (65°C).
- Minimum ambient temperature is −13°F (− 25°C).
- Coil inductance Li is 3.3 mH.

2. Electropneumatic positioners for low power and intrinsic safety. Electropneumatic positioners are intended for automated valves in control applications requiring a milliamp input signal. Standard input ranges are 4 to 20 mA, 4 to 12 mA, and 12 to 20 mA with a supply pressure of 20 to 150 lb/in^2 (2 to 10 bar). They may be used with double-acting or spring-return actuators. Intrinsically safe positioners have an integral signal converter which should be approved as intrinsically safe for class I, division 1, groups C and D and nonincendive for class I, division 2, groups C and D hazardous locations.

3. Hermetically sealed (reed) limit switches. These switches are intended for intrinsically safe or low power circuits where a hermetically sealed switch contact is desirable. These switches are housed in an electrical enclosure and provide open and close indication for pneumatic rotary actuators.

Reed switches are magnetically actuated switches that are encapsulated in a sealed capsule to provide a "pure" signal in low power systems. The sealed housing makes them inherently corrosion resistant.

Typical Reed Switch Specifications

	SPST	SPDT
Make/break distance (in)	0.38/0.5 (10–12 mm)	0.25/0.5 (6–12 mm)
Switching voltage	200 ac/dc	200 ac/dc
Amps-dc (max)		
Switching	0.50	0.25
Carry	1.20	1.20
Contact resistance (Ω)	0.10	0.15
Operating temp.	−40 to 250°F −40 to 121°C	

It must be emphasized that a component with an intrinsically safe approval will only be safe either when installed within the terms of the approval documents and of the relevant codes of practice or when the system configuration has been given prior expert approval.

Typically, the cost of an intrinsically safe component is comparable or slightly higher than the standard explosion-proof device. When one considers that explosion-proof fittings and conduits are no longer needed, on the average the installed system may be less costly for intrinsically safe devices. Also, operating costs are lower as power consumption is negligible. Plus, an intrinsically safe system can be serviced in hazardous areas while the power is on, something which cannot be done with standard explosion-proof devices.

Finally, when an intrinsically safe system is installed correctly, it is always explosion-proof. When standard explosion-proof housings are damaged or there are errors in installation that allow a flame path to develop, the system is no longer explosion-proof. Some insurance companies have rewarded companies who install intrinsically safe systems by charging them lower premiums.

Chapter

12

Installation Practices and Methods

Depending upon where a valve actuator is to be used, it is important that it be able to operate under a variety of adverse environmental conditions. Two to consider are high-temperature conditions and corrosive environments. In addition, the way a valve and actuator are mounted to one another can dramatically affect the performance of the automated valve. This chapter covers this subject.

12.1 Temperature

Generally, the temperature range that a pneumatic actuator will operate in is related to the seals and bearings in the unit. Most will operate in a range of about −4 to 175°F (−20 to 80°C), although optional seal materials can broaden this range.

When pneumatic actuators are used outdoors, where the temperature may fall below freezing, it is often desirable to dry the compressed air to an adequate dew point. Dew point is the temperature at which condensation occurs in air. For such applications, absorption dryers may be used. With these dryers, dew points down to −40°F (−40°C) can be obtained.

Although an actuator may be rated for a specific temperature range, the accessories used may have a temperature limit that falls outside the range. Some solenoid valves, switch boxes, and positioners have components that may not tolerate extreme heat or cold.

With regard to heat, most high-temperature applications are not environmentally generated. When high-temperature fluids are passed through a valve, the heat is transferred through the body of the valve to the actuator. Positioning the actuator below or to the side of the rising heat helps somewhat, but ultimately it is the distance from the pipe (since heat travels through the valve stem and mounting bracket to the actuator) that prevents heat damage.

Installation of actuators in hot line service

When installing an actuator and/or control accessory for valve applications handling extremely hot fluids, it is important to be sure that their maximum working temperatures are not exceeded. This guide for maximum temperatures can be used for the most common control accessories.

Standard pneumatic actuator with Buna seals	175°F (80°C)
With optional Viton seals	212°F (100°C)
With optional Viton seals and optional bronze bearings	250°F (120°C)
Solenoid valve	120°F (50°C)
Positioners	180°F (82°C)
Limit switches	185°F (85°C)

Generally, for actuator mounting, the percent of temperature loss between the fluid temperature and the ambient temperature is dependent on the distance from the surface of the pipe to the bottom of the actuator mounting surface.

The attitude of the valve and the style of the mounting have little effect on the temperature loss gradient. Figure 12.1 gives the percent of temperature loss realized at various distances from the pipe surface.

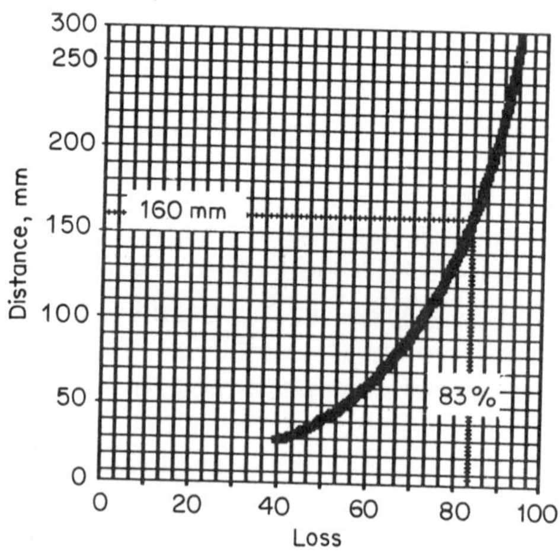

Figure 12.1 Percent of temperature loss realized at various distances from pipe surface.

Procedure

Calculate the percentage temperature loss T_L required using

$$T_L = \frac{T - T_{\text{act}}}{T - T_{\text{amb}}} \times 100$$

where T = temperature of fluid passing through valve
T_{act} = maximum actuator temperature (or lowest maximum temperature for all components)
T_{amb} = surrounding temperature

Example Given T = 600°F (315°C), T_{act} = 175°F (standard) (80°C), and T_{amb} = 86°F (30°C),

$$T_L = \frac{600 - 175}{600 - 86} \times 100 = 83\%$$

From Fig. 12.1, the distance for 83 percent is 6.25 in (160 mm). The mounting hardware between the valve and actuator should be made to provide this distance between the valve surface and the bottom of the actuator.

12.2 Actuator Options for Corrosive Environments

For applications in corrosive environments, valve actuators are available in a variety of materials of construction and coatings. Installing an automated valve with the actuator located vertically above the valve is desirable in corrosive services. If positioned vertically below the valve, an actuator is destined to be harmed by corrosion sooner or later when leaks or dripping occur through the valve stem packing.

Depending upon the manufacturer, standard and optional materials and coatings for the actuator housing are offered in the approximate order of availability, as shown in Table 12.1.

TABLE 12.1 Pneumatic Actuator Body Material and Coating Options

Material	Coating
Aluminum	Anodizing
Steel	Baked enamel
Iron	Epoxy
Plastic	Polyurethane
Stainless steel	

By far, aluminum is the most popular body material for pneumatic and electric rotary actuators. It is lightweight, strong, and economical. It may be cast or extruded, and its composition may be altered for specific conditions. For instance, housing compositions with less than ½ percent of copper are suitable for use offshore or in other salt-spray applications.

Before specifying a valve actuator, it is useful to review the chemical resistance characteristics and other properties of the available coatings or treatments for each. The following chart may be used as a guide.

Chemical Resistance Guide for Coating System and Anodizing Used for Aluminum Actuators

	Resistance*		
Environment	Polyurethane	Epoxy	Anodized
Water			
Fresh	1	1	1
Salt	1	1	1
Solvents			
Alcohols	2	2	2
Aliphatics (M.S, V.M, P, etc.)	1	1	1
Aromatics (xylene, toluene)	1	1	1
Carbon tetrachloride	1	1	1
Ethylene glycol	1	1	1
Ketones	3	3	1
Trichlorethylene	3	4	1
Organic Acids			
Acetic 10%, pH 2.9	2	4	4
Fatty	1	4	7
Citric 10%, pH 2.2	1	1	4
Lactic, dilute, pH 2.4	2	3	3–4
Lactic, concentrated, pH 2.0	3	4	4
Oleic 100%	1	4	4
Oxalic 20%, pH 1.6	1	3	4
Picric 10%	4	4	7
Inorganic Acid			
Hydrochloric 20%, pH 1.5	3	4	4
Hydrochloric 5%, pH 2.5	2	4	4
Phosphoric 65%, pH 1.5	2	4	4
Phosphoric 50%, pH 1.5	1	4	4
Phosphoric 20%, pH 1.4	1	3	4
Sulfuric concentrated (battery acid)	1	4	4
Sulfuric 10–50%, pH 1.0	4	4	4
Sulfuric 10%, pH 1.2	1	4	4
Boric (saturated solution), pH 5.2	1	1	3

Chemical Resistance Guide for Coating System and Anodizing Used for Aluminum Actuators (*Continued*)

Environment	Resistance*		
	Polyurethane	Epoxy	Anodized
Acid Salts			
Aluminum nitrate 10%	1	1	3
Ammonium chloride	1	1	4
Copper sulfate	1	1	4
Ferric nitrate	1	4	4
Zinc sulfate	1	1	3–4
Alkalis			
Ammonium hydroxide 28%, pH 11.4	1	1	4
Ammonium hydroxide, dilute, pH 10.7	1	1	4
Calcium hydroxide (lime), pH 12.4	1	1	4
Potassium hydroxide 50%, pH 14	1	1	4
Potassium hydroxide 25%, pH 13	1	1	4
Potassium hydroxide 10%, pH 14	1	1	4
Sodium hydroxide 50%, pH 14	1	1	4
Sodium hydroxide 20%, pH 13	1	1	4
Sodium hydroxide 5%, pH 12	1	1	4
Sodium hypochlorite (bleach)	1	1	4
Alkaline Salts			
Barium sulfide	1	1	3
Sodium bicarbonate	1	1	4
Sodium carbonate	1	1	1
Sodium sulfide	1	1	3
Trisodium phosphate 10%	1	1	3
Gases			
Ammonia	2	2	4
Chlorine, dry	1	1	2
Hydrogen chloride	2	2	4
Hydrogen sulfide, wet	1	1	3
Sulfur dioxide, dry	1	1	3
Sulfur dioxide, wet	2	2	4
Chloride, wet	2	2	3
Miscellaneous Organic Materials			
Cutting oils	1	1	1
Detergents	1	1	1
Gasoline (unleaded/regular)	1	1	1
Hydraulic fluid	3	3	3–4
Lubricating oils	1	1	1
Refinery crudes	1	1	3
Sewage	1	1	3
Skydrol 500	1	1	?
Sugar liquids	1	1	2
Waste sludges	1	1	?

Chemical Resistance Guide for Coating System and Anodizing Used for Aluminum Actuators (*Continued*)

Environment	Resistance*		
	Polyurethane	Epoxy	Anodized
Miscellaneous Exposure			
General weathering	1	1	1
Industrial fumes; salt atmosphere	1	1	1
Mold; mildew	1	1	1
Light abrasion	1	1	1
Heavy abrasion	1	1	2

*Resistance: 1—excellent, 2—good, 3—fair, 4—not recommended.

Coatings for actuators and other process equipment are usually evaluated based on abrasion, adhesion, and impact. Following is a typical comparison of two common coating materials.

Comparing Polyurethane Coating Systems to a Standard Epoxy Finish

Tests	Polyurethane	Epoxy
1. Abrasion resistance*	103 mg, excellent	160 mg, good
2. Exterior durability	Excellent (no chalking)	Very good (with chalking)
3. Pencil hardness	2H	F
4. Elcometer adhesion[dg]	460 lb/in^2 to break	295 lb/in^2 to break
5. Dry heat resistance	200°F (93°C)	200°F (93°C)
6. Moisture condensation resistance	100°F (37°C), 120 h, no failure	100°F (37°C), 120 h, limited
7. Salt fog resistance	1000 h (good)	1000 h (good)
8. Direct impact resistance‡	160 in · lb (18 N · m)	15 in · lb (2 N · m)
9. Scrub resistance	10,000 cycles; no gloss change	10,000 cycles; no gloss change

*Abrasion resistance test—abrasive wheel is applied to finished surface for 1000 cycles. Published is the total weight of particles removed, in milligrams.

†Elcometer adhesion test—measures the amount of pressure required to remove coating from finished surface.

‡Direct impact resistance test—various weights are dropped from a measured distance to determine impact (chip resistance) that surface will withstand without cracking.

12.3 Installation and Operation

Two rules of thumb for actuator mounting and operation are that (1) it be mounted so that its length is parallel to the run of the pipeline and (2) it operate in a quadrant such that counterclockwise rotation will open the valve and clockwise rotation will close it (see Fig. 12.2).

When supplied with a factory-installed solenoid pilot valve, a pneumatic actuator operates counterclockwise (to open) when the solenoid is energized and clockwise (to close) when the solenoid is deenergized. The reverse of this is available but should be specified.

Some actuator manufacturers have gone to the modulator concept

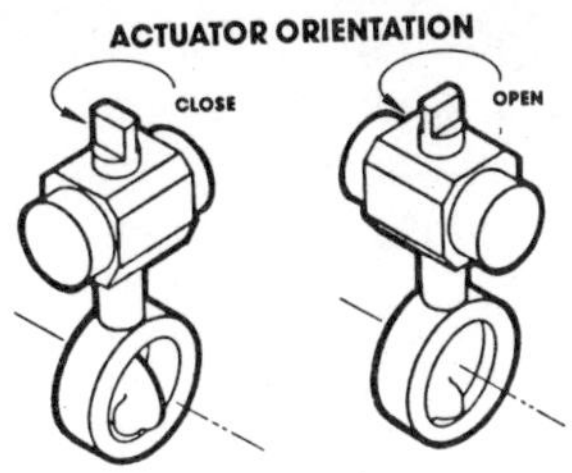

Figure 12.2 Typical pneumatic actuator mounting and operation. (*Courtesy of Demco Division of Cooper Industries.*)

for solenoid valve mounting by integrating a manifold in the actuator body. This allows the operation of the valve to be reversed in the field by inverting it. Most electric rotary actuators are supplied with wiring diagrams and are assembled to comply with this practice of open-close directional operation.

Actuator mounting hardware design

An automated valve package is constructed of three major components: the valve and actuator, of course, are two, but an element often underestimated in importance is the mounting hardware. Improperly designed mounting hardware is dangerous, reduces the performance life of the package, may cause premature seal failure and valve leakage, and if made from improper materials, may corrode and fail well before expected. Let's review these situations individually.

Any product that does not perform as expected may become a potential hazard to operating personnel. For example, some actuators have a flat visual position indicator at the top and bottom output shaft areas. Dimensional drawings do not always show that the indicator is present, which means that mounting hardware may not be designed with adequate clearance for the indicator which is fixed to and rotates with the shaft. When mounted, the actuator and mounting bracket can sandwich the indicator and prevent free rotation of the shaft. In a very short time this can cause a seizure which may prevent a valve from closing fully. Hazardous fluids may leak through the valve and represent imminent danger.

A similar example is a bracket that is too short (or a coupling that is too long). If, when initially assembled, there is a gap of as little as an eighth of an inch, the tendency is for the assembler to use the mounting bolts to pull the surfaces together. This is particularly a problem for actuators with tapered shafts that are bottom-loaded. Taking up the surface gap causes the coupling to force the actuator shaft further into the actuator body. Because it is tapered the shaft wedges itself into the body and soon becomes very difficult to operate. As the actuator—particularly a spring-return actuator—is using an unexpected amount of its output torque to overcome the additional wedging forces, there may not be enough torque left to fully seat the valve. Again this is another potentially hazardous situation.

Another condition in which excessive torque and premature seal wear may be caused by the mounting hardware is when the top and bottom mounting surfaces of the bracket are not concentrically aligned. Misalignment puts eccentric loads on the actuator and valve shafts and requires additional torque for operation. One solution is to specify that mounting brackets be machined from one side so that all bolt patterns are drilled from the same point of reference.

Because mounting brackets are usually made from square tubing the top and bottom surfaces should be machined flat and parallel. Eccentricity may also be caused by an improperly machined drive coupling. The valve and actuator ends of the coupling must be machined on a common centerline to avoid this.

For automated valves intended for high cycle or modulating control applications the machined parts of the coupling should be specified to tight tolerances to eliminate slop. Couplings between the actuator auxiliary drive and positioner input shaft should have close tolerances plus a set screw to eliminate any lost motion. When installed in highly corrosive areas, mounting hardware should be made from appropriate materials or specially treated. Stainless-steel mounting kits are common in many areas.

Equally important as mounting hardware design is the provision for actuator mounting that valve manufacturers provide. Today automated valve users are specifying that pressure-retaining bolts not be used for actuator mounting. This includes bonnet bolts on top-entry valves and body bolts on two- and three-piece ball valves. Some engineers will accept a double-mounted bolt (see Fig. 12.3). For a top-entry design this would mean replacing the bonnet bolts with larger studs. One nut goes onto each stud to hold down the bonnet, and then the mounting bracket is placed onto them and fastened down using the balance of the length of the stud.

Some valve manufacturers have addressed this issue by providing predrilled actuator mounting surfaces that do not employ any of the pressure-containing bolts. Two examples are flange mounts and center body mounts. The flange-mount design is usually provided on top-entry valves (see Fig. 12.4). Each flange is flatted on top and predrilled. The mounting bracket spans across the flanges and becomes a robust, safe design.

Some three-piece ball valves have a drilled and tapped mounting surface on the center section to eliminate the use of the body assembly

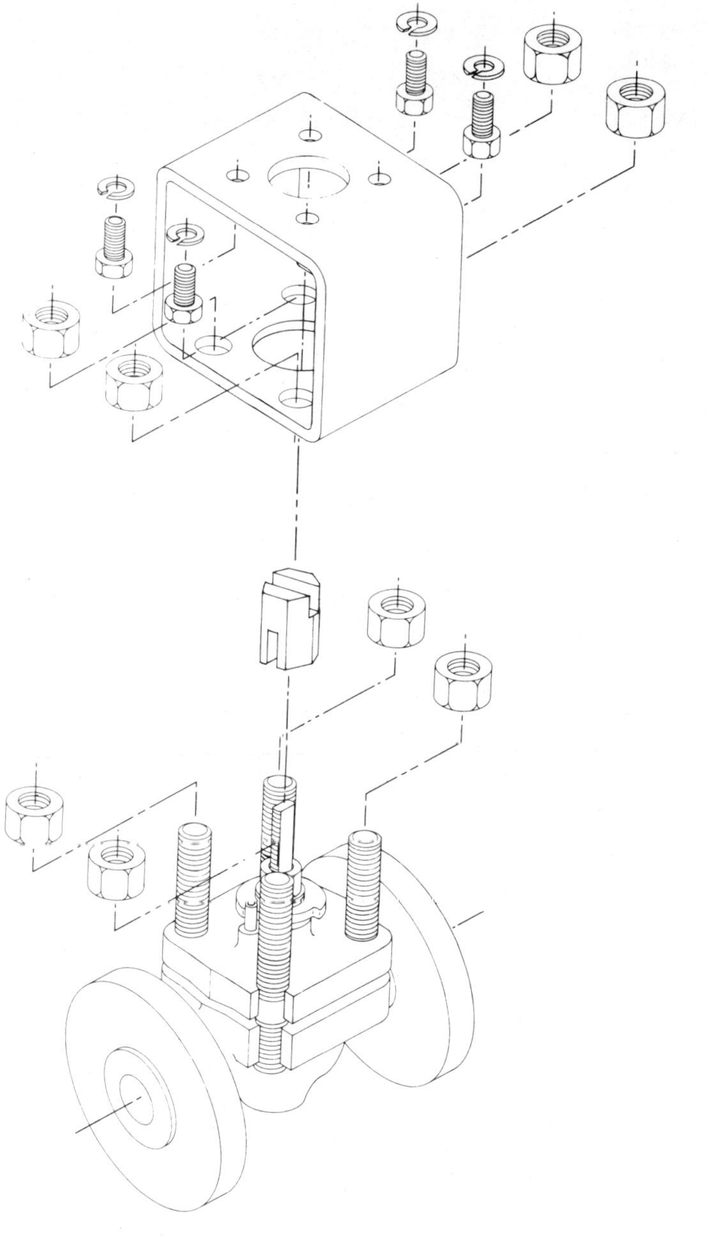

Figure 12.3 Example of double-nut bonnet mount for actuators. (*Courtesy of ITT Engineered Valves.*)

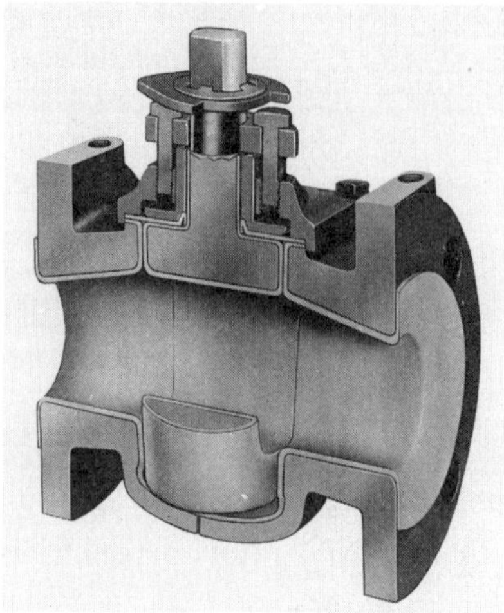

Figure 12.4 Flange-mount design uses tapped holes on flatted tops of the valve flanges. (*Courtesy of The Duriron Company.*)

bolts (see Fig. 12.5). The most important reason for using these safer designs is that in the event that an actuator needs to be removed from the valve, it is never necessary to destroy the integrity of the manufactured valve components or subject maintenance personnel to the potential hazard of cracking a valve body seal that is under pressure.

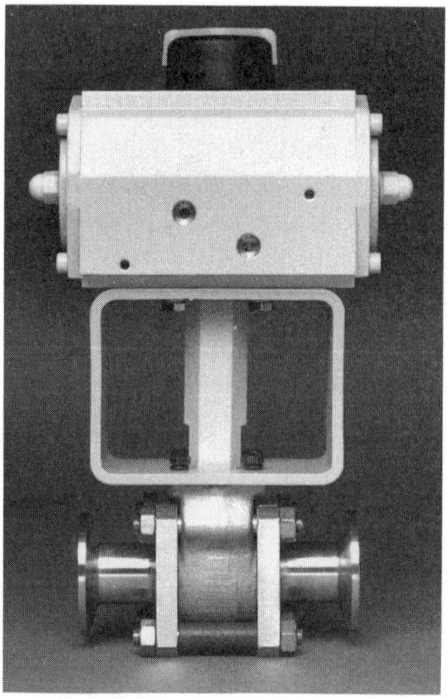

Figure 12.5 Three-piece ball valve with tapped center body for actuator mounting. (*Courtesy of MCF Valves.*)

One common problem for users of automated valves who prefer to select the valve and actuator individually and create an ideal package is mounting hardware availability. Because there are currently no standard actuator or valve mounting dimensions each new pairing of valve and actuator often requires the design of an entirely new and unique mounting kit.

Some manufacturer's organizations including the Manufacturers Standardization Society (MSS) have adopted in whole or in part the intent of the International Standards Organization (ISO) committee's standard 5211. This standard specifies the bolt pattern and drive dimensions for actuators and for the corresponding valve size by torque. Many European butterfly valve manufacturers have adopted standard 5211 as just about every actuator manufacturer has. Actuators manufactured to ISO 5211 can be mounted directly, without the need for a mounting kit, to any butterfly valve with the same pattern. Even for valves that cannot be directly mounted, for example, a ball valve where packing adjustment must be accessible, a standardized mounting pattern makes mounting kit production and availability that much better.

ISO 5211 specifies the bolt circle (flange) and drive shaft dimension for the bottom of the actuator. It specifies the maximum torque that may be transmitted through the mounting flange couplings and takes into account bolt material, allowable bolt stress, and the coefficient of friction between the mounting flanges. For the most part automated valve users in Europe have driven the adoption of these standards by manufacturers by refusing to install equipment that does not comply with the ISO 5211 guidelines.

Standards of this type are also being developed for pilot valve manifold mounting and auxiliary equipment mounting surfaces on the tops of actuators. This trend not only improves mounting hardware design but also expands the availability of products that may be used interchangeably to automate a valve.

In some installations, particularly with large actuators, there may be a condition where the actuator is mounted parallel to the pipeline but interferes with another piece of equipment, such as a tank, which cannot be moved to accommodate the actuator's space requirement. Some actuators are designed with a square valve mounting pattern (ISO 5211) or bolt circle. This allows it to be mounted 90° from the normal mounting position, or perpendicular to the pipeline. In many cases, this solves the problem.

Valve installations

A rotary stem valve such as a butterfly valve should preferably be mounted with the valve stem horizontal, to prevent the bottom bear-

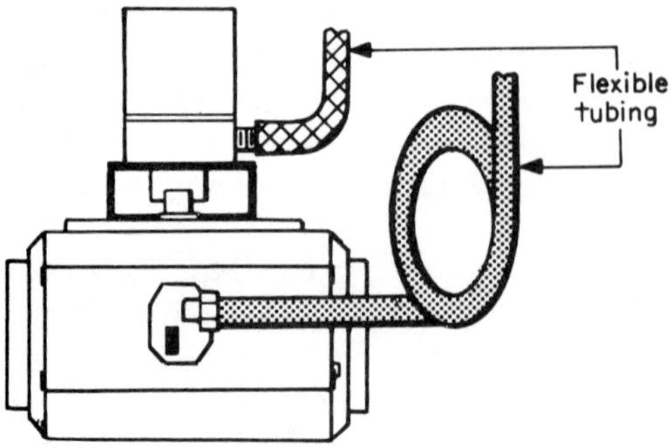

Figure 12.6 Flexible tubing and conduit is often specified in high-vibration installations.

ing from being at the low point in the pipe, where dirt will collect. The direction of opening of ball or butterfly valves should be such that any solids that have settled out of the stream while the valve was closed will be swept along as the valve opens.

In vertical pipeline installations, a rotary valve with a compact actuator has several advantages over a linear-motion valve. For instance,

Low-profile actuators usually do not require external supports.

Actuators that are short and compact are not subject to severe vibration damage.

The initial cost is less.

The installation cost is less.

If an automated valve is subject to vibration when put into service, there are some potential problems that can occur, depending on the severity of vibration. Fittings may loosen, stress damage may occur, control accessories may require frequent adjustments, and conduit and air line fittings may develop cracks or leaks.

For this reason, it is important that air tubing and electrical connections to the actuator be sufficiently flexible (see Fig. 12.6). Plastic air tubing is inherently flexible. Copper or stainless-steel air tubing should be coiled (one or two large loops) at the actuator. Flexible electric conduit may be used at the switch box, solenoid valve, etc., but it is important to be aware of electrical codes and requirements related to area and device classifications.

12.4 Retrofitting

As plant engineers become more aware of the many advantages of plantwide automation, they are increasingly faced with cost, downtime, safety, and any number of factors that plant upgrading represents. From a simple liquid-level control system to an entire process control center, entering into a plant automation upgrade program requires a great deal of funding and time.

Automated valves in a new process

Automated valves are often required in proposed process control system upgrades to operate in conjunction with any new controllers, control stations, or safety systems.

The plant engineer may not be aware that it is not always necessary to replace manual valves when automated valves are required and that existing manual valves may be fitted with actuators in the field. This is known as actuator retrofitting.

What to consider

To thoroughly evaluate the savings that may be realized when considering an actuator retrofit, let us review some important points. When a valve is removed from a line, it is usually necessary to purge the piping system, shut down production, and remove and reinstall the insulation. If valves are welded into the line, retrofitting may involve the additional cost of cutting the lines and providing new or modified pipe ends to adapt to the new valve. Many times an existing valve may be in good condition and may not need to be removed and discarded.

Additionally, there is the concern about worker-hours, dimensionally interchangeable valves, and coordinating the arrival of new valves with a scheduled shutdown. This may add to the overall burden of valve replacement and system upgrade.

Furthermore, plants that, over the years, have installed new valves where and when necessary during normal maintenance may find that when plantwide automation is required, there are many valves for a similar service that may vary by

1. Manufacturer
2. Valve type
3. Size
4. Pressure rating
5. Temperature rating
6. Torque

The job of retrofitting an actuator to these valves may not seem simple.

However, there are still significant savings to be realized and important reasons to continue with an evaluation of actuator retrofitting.

Selecting an actuator

The first step in the retrofit process requires the selection of an actuator type and manufacturer with a complete enough product line to satisfy a wide variety of applications. Another factor to consider when selecting an actuator for a retrofitting program is if it is available with a complete line of control accessories: switches, pilot valves, positioners, and, most important, mounting hardware.

Independent actuator manufacturers

If indeed the process area targeted for upgrade contains valves of many manufacturers and many valve types (ball, plug, butterfly), it may be impractical and inefficient to attempt to contact each valve manufacturer individually and then to evaluate the sometimes conflicting recommendations each one makes.

In this case, an independent actuator manufacturer is a good choice. Independent actuator manufacturers do not market valves but provide actuators for any rotary valve. Their business is to provide a single source for valve automation systems. Their expertise is in manufacturing actuators for any and all valves in a process system and providing a single responsible source for valve automation. Many manufacturers, for instance, have a valve automation program that deals specifically with retrofit applications. A factory-trained representative makes a step-by-step evaluation of the system and the valves in that system, and then provides

Location audit: to determine valve quantities, types, and mounting arrangements

Actuator sizing: to be based on available air pressure, required operation, and valve torque and service

Accessory selection: to develop a complete control package that matches the proposed or existing system requirements and maintains strict product integrity

Complete submittal package: to include technical drawings, data sheets, signal, power, and performance specifications, and wiring diagrams

Scheduling: to provide valve automation packages when required for convenient installation during a shutdown period

A factory representative should also be prepared to do a dimensional

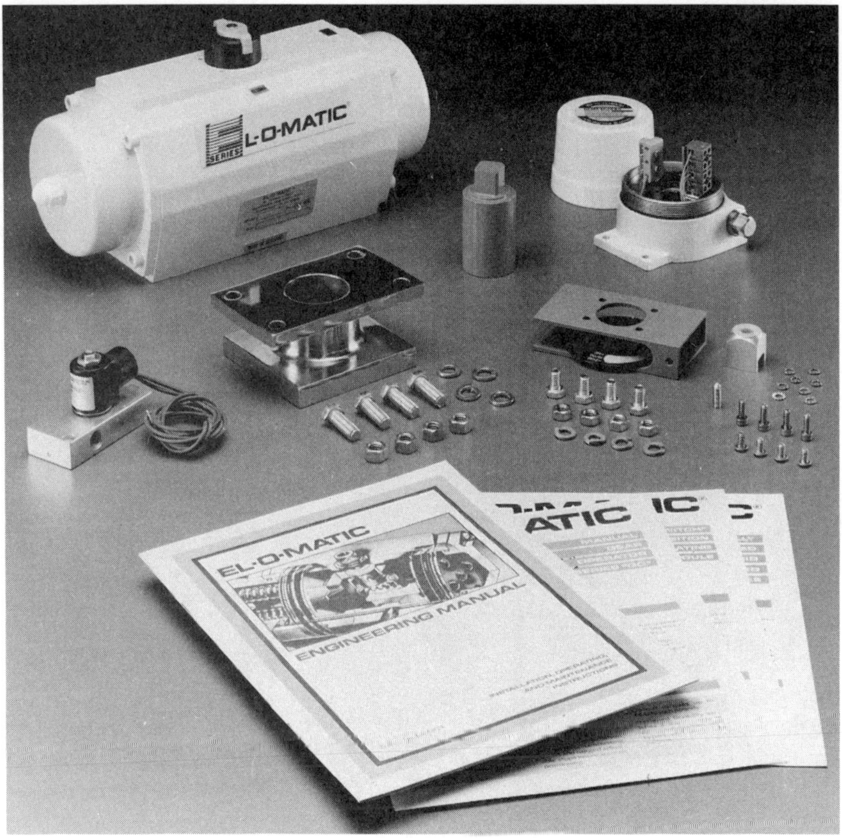

Figure 12.7 Typical retrofit kit includes actuator, accessories, mounting kit, and manuals for field automation of existing valves. (*Courtesy of El-o-matic, Inc.*)

"take off" of individual valve bonnets for the design and production of mounting hardware. The representative may then provide the engineer with a retrofit kit (see Fig. 12.7) that includes mounting hardware, control accessories, data sheets, and an installation manual, prepackaged and labeled for convenience and simplicity.

Retrofitting actuators and accessories to manual valves is an efficient method for plant upgrading. Shipments may be scheduled to spread the work over a period of time. In many cases, this minimizes downtime since some work may be performed during a scheduled shutdown and without costly, time-consuming piping disruptions. A thorough audit of valve locations and the surrounding equipment reduces complications when the equipment is ready for installation. There are significant savings to be realized when existing valves that do not require replacement remain in service.

Appendix

A

Reference Guides

BUTTERFLY VALVE MATERIAL SELECTION GUIDE

The following application data is for use in determining the available material *most* likely to give satisfactory service and is not meant to imply any type warranty. However, certain factors such as temperature, concentration and combinations of chemicals may affect the fluid resistance of any material. In questionable applications, testing is always the most positive way to determine suitability.

Some materials are available in certain valve sizes and series only. Refer to the specific valve sections for availability.

APPLICATION CODE: E — Excellent
G — Good
F — Fair
U — Unsatisfactory

Stem & Disc Materials **Seat Materials**

	Nickel Plated Ductile Iron	416SS	316SS	Monel	Manganeze Bronze	Aluminum Bronze	Lead	Alloy 20	Hastelloy C	PVF	Buna N	Neoprene	Hypalon	Viton	EPT	Natural Rubber
Acetaldehyde	U	U	E	E	U	U	E	E	E	G	G	F	U	F	E	U
Acetic Acid – Crude	U	U	E	F	U	U	F	E	E		G	E	E			
Acetic Acid – Pure	U	U	E	F	F	F	F	E	E	E	G	E	E			
Acetic Acid – 10%	U	U	E	F	U	U	F	E	E		G	E	E			
Acetic Acid – 80%	U	U	E	F	U	U	F	E	E		G	E	E			
Acetic Acid – Anhydride	U	U	E	F	U	U	F	E	E	U	U	E	E	U	F	F
Acetone	G	G	E	E	E	E	G	E	E	U	U	U	U	U	E	F
Acetophenone	U	U	G	E	U	U	E			F	U				E	
Acetylene	G	E	E	E	E	E	G	E	U		G	F	F	E		F
Acrylonitrile	G	G	E	E	E	E	E			U	U	F	F	U	U	U
Air (Dry)	E	E	E	E	E	E					E	E	E	E	E	
Alcohol – Amyl	F	G	E	E	E	E	G	E		E	F	F	G	G	E	G
Alcohol – Butyl	F	G	E	E	E	E	G	E	E	E	F	G	G	E	F	F
Alcohol – Ethyl	U		E	E	E	E	G	E	E	E	G	G	G	G	G	G
Alcohol – Methyl	U		E	E	E	E	G	E	E		F	E	E	F	E	G
Alum – Ammonium	U		G								G	G	F	G		F
Alum – Chrome	U		G								G	G	G	G		F
Alum – Potassium	U	G				G					G					
Alumina	G	G	G	G	G	G	U	E	E		E	F	G	G	E	E
Aluminum Chloride	U	U	F	G	U	U	U	E		E	G	G	G	E	E	G
Aluminum Fluoride	U		G	G			G	E		E	G	G	G	G	E	G
Aluminum Hydroxide	U		G	G	G	G	G			E	G	G		G	G	
Aluminum Sulphate	U	G	G		U	U	U	G	E		E	E	E	E	E	E
Amines	U	F	E	G	G			E	U		U	U	U	U	F	
Ammonia, Anhydrous	F		E	E	U	U	U		E	E	G	G	U	U	G	U
Ammonia Solutions	F	G	E	G	U	U	E	E	E		G	G	U	U	G	U
Ammonium Chloride 50% 180°F	U	F	G	G	U	U	F	E	E	E	E	E	E	E	E	E
Ammonium Hydroxide	F	G	E	F	U	U	U	E	E	E	U	E	E	G	E	U
Ammonium Nitrate 5% 60°F	F	G	E	U	U	U	F	E	E	E	E	E	G	E	E	G
Ammonium Phosphate	U	G	E	G	U	U	U	E	E	E	E	E	E	E	E	G
Ammonium Sulphate 90% 180°F	U	F	G	G	U	U	U	G	G	E	E	E	E	E	E	G
Amyl Acetate	F	G	E	G	G	G	E	E	U	F	U	U	U	U	G	U
Amyl Chloride	F	G	E	G	F	E		E	E	E	U	U	U		U	U
Aniline 90% 70°F	F	G	G	G	U	U	E	E	E	F	U	U	U	F	F	U
Aniline Dyes	F	G	E	E	F	F	E	E	E		U	F	F	G		
Antimony Chloride	U			G			E		E		G	G	G	E		
Arsenic Acid	U	G	G	G	U	U	U	E		E	E	E	E	E	G	G
ASTM Oil #1	F	E	E	E	E	E	G	E	E		E	G	G	E	U	U
ASTM Oil #3	F	E	E	E	E	E	G	E	E		E	U	U	E	U	U
ASTM Ref. Fuel A	F	E	E	E	G	G	G	E	E		E	G	G	E	U	U
ASTM Ref. Fuel B	F	E	E	E	G	G	G	E	E		G	F	F	E	U	U
ASTM Ref. Fuel C	F	E	E	E	G	G	G	E	E		G	F	F	E	U	U
Asphalt	E	E	E	E	E	E	E	E	E		G			E	U	U
Barium Carbonate 60°F	U		G	G	G	G			E	E	E	E			E	
Barium Chloride	U		G	G	G	F	G		E	E	E	E	E	E	E	E
Barium Hydroxide	F	E	E	G	U	U		E		E	E	E	E	E	E	G
Barium Sulphate 60°F	U		E	G	G	G				E	E	E			E	
Barium Sulphide	F	E	G	F	U	U	U	E		E	E	E	U			U
Beer (Beverage)	U	E	E	E	U	U	U	E	E	E	G	G	E	E	E	G
Beet Sugar Liquors	F	E	E	E	F	F	U	E			E	E	G	E		G
Benzaldehyde	F	E	E	G	E	E	U	E		E	U	U	U	U	E	U
Benzene (Benzol) 70°F	F	E	E	G	E	E	E	E	E	G	U	U	U	G	U	U

Figure A-1 Valve material selection guide. (*Courtesy of Demco division of Cooper Ind.*)

BUTTERFLY VALVE MATERIAL SELECTION GUIDE

APPLICATION CODE: **E**— Excellent
G— Good
F— Fair
U— Unsatisfactory

	Stem & Disc Materials										Seat Materials					
	Nickel Plated Ductile Iron	416SS	316SS	Monel	Manganeze Bronze	Aluminum Bronze	Lead	Alloy 20	Hastelloy C	PVF	Buna N	Neoprene	Hypalon	Viton	EPT	Natural Rubber
Benzoic Acid 5% 70°F	U	G	E	G	G	G	U	E	E	E	U	U	U	E	U	U
Black Sulphate Liquor	F	G	E		F	F		E			G	G	G	E	G	F
Bleaching Powder-Wet	U	G	G		U	U		E			G	E	F			F
Borax (Sodium Borate)	U	F	E	E	U	U		E	E	E	G	E	E	E	E	G
Boric Acid 5% 200°F	U	F	E	G	F	F	F	E	E	E	E	E	E	E	E	G
Brine (Acid)	U		G	G	G	G	G	·		E	E	G	G		E	U
Brine-Chlorinated			U	G	G	G								G	G	
Bromine-Gas	U	U	U	F	G		E	U	E	E	U	U	U	E	U	U
Bromine-Water	U	U	U	F			E	U	E	E	U	G	G	E	U	U
Butadiene	F	G	E	E	G	G	E		E	E	F		G	G		U
Butane-Butylene	G	E	E	E	E	E	E	E	E	E	G	G	G	E	U	U
Butyl Acetate	G	E	E	E	E	E	E	E	E	U	U	U	U	U	U	U
Butyric Acid 5% 70°F	U	G	E	G	F	G	U	E	E		U	U	U	G	G	U
Calcium Bisulfite	U	G	E	U	U	U	E	G		E	E	E	E	E	U	U
Calcium Carbonate 60°F	F		G	E	G	G	E		E	E	E	E		E		
Calcium Chlorate 20% 160°F		G	E	G	U	U		E	E	E		E	E			
Calcium Chloride	F	G	G	G	F	E	U	G	E	E	E	E	E			
Calcium Hydroxide 50% 175°F	F	E	E	E	U	U	G	E	E	E	G	E	E	E		
Calcium Hypochloride		G	G				G	G			U	U	E	E		U
Calcium Sulphate 90% 60°F	F	E	E	G	E	E		E		E	F	F		E		
Calgon	F		E					E			E	E		E		
Caliche Liquor	U		E					E			E	E	E	E	E	E
Cane Sugar Liquors	F	E	E	G	E	E		E			E	E				
Carbon Dioxide (Dry)	F	E	E		E	E		E		E	E	G	G			G
Carbon Disulfide	U		E	G	G	G	E			G	U	U	U	E	U	U
Carbon Tetrachloride (Dry)	U	G	G	E	G	G	F	E		E	E			E		
Carbonic Acid	U	G	E	E	F		G	E	E	E	G	E	E	E	E	E
Castor Oil	G	G	E	E	G	G	G	E	E	E	G	E	E	E	G	G
Caustic Solutions 34% 200°F	U	G	E	F	U	U		F	F		F	F	F	E	G	U
China Wood Oil (Tung)	F		F	F	F	F		E	E		E	E	G	E	U	U
Chlorine Gas-Dry 70°F	U	F	G	G	G	G	U	G	E	E	U	U	G	E	F	U
Chlorobenzene 90% 70°F	F	E	G	G	G	G		E		E	U	U	U	G	U	U
Chloroform 70°F	F	G	E	E	G	G	E	E		E	U	U	U	E	U	U
Chlorosulfonic Acid 10%	U	U	G	F	U	G	E	U	E	F	U	U	U	U	U	U
Chocolate											E	U				
Chromic Acid 5% 70°F	U	G	G	F	U	U	U	E		E	U	U	E	E	U	U
Citric Acid 5% 150°F	U	F	E	G	F	F	E	E	E	E	E	E	E	E	E	F
Coconut Oil (Food)	U	U	E	G	G	G	U	E		E	E	F	F	E	F	U
Coffee (Food)	U	U	E	G	F	F		E		E	E	E	E	E	E	E
Copper Chromate			E													
Copper Sulphate 80% 175°F	F	G	E	F	U	U	E	E		E	E	E	E	E	E	E
Corn Syrup			E												E	
Cottonseed Oil	G	E	E	E	E	E		E	E	E	E	F	G	G	F	U
Creosol	U	F	E	E				E		F	U	U	U	F	U	U
Creosote	E	E	E	E	G	G	E	E	E		G	U	F	E	U	U
Cresylic Acid	U	G	E	G	F	G		E		G	U	U	U	E	U	U
Crude Oil	G	G	E	E	E	E	E	E	E		E			E	U	U
Cyclohexane	F	E	E	G	G	G		E		E	U	U	U	E	U	U
Dextrose (Food)	U	U	E					E			E					
Diacetone	U		E		E	E				U	U	U	U	U	E	U
Diamylamine	F	E	E		E	E		E			G	F		U		G
Dichlorethene	U		G	G	F	G		G		F	U	U		G	U	
Diesel Fuels	E	E	E	E	G	G		E	E	E	E	F	F	E	U	F
Diethyl Amine	F	E	E	G	E	E		E		U	G	F	F	U	F	U
Dimethane Sulfide													F	U	G	
Dowtherms	G	E	E	E	F	G		E			U	G	G	E		
Drilling Mud	G		E	E	G	G					E	F	E	E		G
Ethers	U		G	G	G	G	E	E	E	F	U	U	U	F	F	U
Ethyl Acetate	F	G	E	E	G	G		E	E	F	U	U	U	U	G	U
Ethyl Chloride 5% 60°F	F	G	E	G	F	E		E	E	E	E	G	G	G	G	G
Ethylene Dichloride	U		G	G	G	G	E			E	U	U	U	F	U	
Ethyl Glycol	G	E	E	E	E	E	E	E	E	E	E	G	E	E	E	E

Figure A-1 *(Continued)*

BUTTERFLY VALVE MATERIAL SELECTION GUIDE

APPLICATION CODE: **E** — Excellent
G — Good
F — Fair
U — Unsatisfactory

Stem & Disc Materials **Seat Materials**

	Nickel Plated Ductile Iron	416SS	316SS	Monel	Manganeze Bronze	Aluminum Bronze	Lead	Alloy 20	Hastelloy C	PVF	Buna N	Neoprene	Hypalon	Viton	EPT	Natural Rubber
Ethylene Oxide	G	G	G	G	U	U		E		U	U	U	U	U	F	U
Ferric Chloride	U	U	U	U	U	U	U		U	E	E	G	G	E	E	E
Ferric Nitrate (pH7+) 5% 60°F	U		G	F	U	U		E		E	E	E	E	E	E	E
Ferric Sulphate 5% 60°F	U		E	E	U	U				E	E	E	E	E	E	E
Ferrous Sulphate	F	G	G		U	U	E	E		E	E	E		E	E	E
Fluorine 70°F	U	U	F	G	U	U	G	G		E	G	F	G	G	F	
Fluosilicic Acid	U	F	G	G	F	F		E		E	E	G	E			G
Formaldahyde 70°F	U	G	E	G	E	E		E	E	G	F		E	E		F
Formic Acid 5% 150°F	U	G	E	G	G	G	E	E	E	E	G	E	E	E	E	E
Freon 11	F	E	E	E	E	E	E	E		E	G	G	E	G	U	
Freon 12	F	E	E	E	E	E	E	E		E		G	U	U	U	
Freon 22	F	E	E	E	E	E	E	E		E	U		E	U	E	
Freon 113	F	E	E	E	E	E	E	E			G	G	E	G	U	U
Freon 114	F	E	E	E	E	E	E	E			G					
Fructose			E												E	
Fruit Juices (Food) 70°F	U	U	E	E	U	U		E	E		E	E	F	E	G	G
Fuel Oil	F	E	E	E	E	E	E	E	E	E	G	F	F	E	U	U
Furfural	F	E	E	G	G	G		E	E	U	U	U	U	U	F	U
Gallic Acid 5% 200°F	U		E	G				E	E	F	G	G	G	E	G	E
Gasohol	F	G	E		G	G				E	U			G	U	
Gasoline — Regular	G	E	E	E	E	E		E	E	E	G		G	E	U	U
Gasoline — Unleaded	F	E	E	E	E	E				E	U			E	U	
Gelatin (Food)	U	U	E	G	G	G		E			E	E	E	E	E	E
Glucose	U	F	E		E	E		E		E	E	E	E	E	E	E
Glycerine/Glycerol 70°F	F	E	E	G	G	G		E	E	E	E	E	E	E	E	E
Heptane	F	G	E	G	E	E		E		E	E	G	G	E	U	U
Hexane	F	G	E	G	E	E		E		E	E	G	G	E	U	U
Hydraulic Oils	F	G	E	E	E	E		E			G	F	E	E	U	U
Hydrobromic Acid 200°F	U	U	U	U	U	U	G	U	E	E	U	U	E	E	E	E
Hydrochloric Acid 10% 60°F	U	U	U	F	U	U	G	U	E	E	F	G	G	E	G	
Hydrochloric Acid 20% 60°F	U	U	U	U	U	U	G	U	E	E	F	G	G	E		
Hydrocloric Acid 35% 60°F	U	U	U	U	U	U	F	U	G	G	F	G	G	E		
Hydrocyanic Acid	U		G	F	U	U	U	E	E	E	G	G	E	E	G	G
Hydrofluoric Acid 48%	U	F	U	F	U	U	U	F	E	E	U	U	E	E		
Hydrofluoric Acid 60%	U	U	U	U	U	U	U	F	E	E	G	U	G	G	E	G
Hydrofluoric Acid 100%	U	U	U	U	U	U	U	F	E	E	U	U	U	F		
Hydrofluosilicic Acid 5% 70°F	U	G	G	G	F	E		G	E	G	G	G	E	E	E	E
Hydrogen	F	G	E	E	F	F	E	E		E	E	E	E	E	E	G
Hydrogen Peroxide 90%	U	F	G	G	U	U	U	G	E	E	U	U	G	G	F	U
Hydrogen Sulfide — Dry	F	F	E	G	F	F	E	G	E	E	F		G	G	E	G
Hydrogen Sulfide — Wet	F	F	G	F	F	F	G	E		E			G	G	G	U
Iodine Solution	U	U	U	U	U	U	E	U	E	E	F	U	G	E	G	F
Iso-Octane	F	E	E	E	E	E		E	E	E	E	G	G	E	U	U
Isopropyl Alcohol	F	E	E	E	E	E		E	E		G	F	E	E	E	E
Isopropyl Ether	F	E	E	E	E	E		E	E		G	F	F	U	U	U
Kerosene	E	E	E	E	E	E		E	E	E	E	G	U	E	U	U
Lacquer Solvents	F	E	E	E	E	E		E	E		U	U	U			
Lactic Acid 5% 70°F	U	F	G	G	U	U	E	E	E	U	G	E	E	E	G	G
Lard Oil 70°F	G		E	G	G	G		E	E	E	E	F	U	E	U	U
Lemon Oil	U	F	E	E	G	E	U	E	E	E	G	F				
Linseed Oil	F	E	E	E	E	E	E	E	E	E	E	G	G	E	U	
Lubricating Oil	E	E	E	E	E	E	E	E	E	E	E	F	G	E	U	U
Magnesium Chloride 4% 75°F	F	F	G	G	F	F	U	E	E	E	E	E	E	E	E	E
Magnesium Hydroxide	F	E	E	G	G	G		E		E	G	E	E	E	E	G
Magnesium Sulphate 5% 120°F	F	G	E	E	E	E		E	E	E	E	E	E	E	E	G
Mecuric Chloride 3% 60°F	U	U	F	U	U	U		F	E	E	E	E	E	E	E	
Mecuric Cyanide	U		E	U	U	U		E	E	E	F	F			G	
Mercurous Nitrate (pH7+)	U		G	U	U	U		E	E	E	F	F				
Mercury	E	E	E	E	U	U	F	E	E	E	E	E	E	E	E	E

Figure A-1 *(Continued)*

BUTTERFLY VALVE MATERIAL SELECTION GUIDE

APPLICATION CODE: **E** — Excellent
G — Good
F — Fair
U — Unsatisfactory

Stem & Disc Materials **Seat Materials**

	Nickel Plated Ductile Iron	416SS	316SS	Monel	Manganeze Bronze	Aluminum Bronze	Lead	Alloy 20	Hastelloy C	PVF	Buna N	Neoprene	Hypalon	Viton	EPT	Natural Rubber
Methyl Acetate	F	G	E					E			U	U	U	U	G	U
Methyl Acetone	F	E	E		E	E		E			U	U	U			U
Methyl Cellosolve	F	E	E	E	E	E		E		E	U	U	U	U	G	U
Methyl Chloride	G	G	E	G	F	E	E	E		E	U	U	U	G	F	U
Methyl Ethyl Ketone (MEK)	E	E	E		E	E	E	E		U	U	U	U	U	E	U
Milk (Food)	U	U	E	F	U	U		E	E	E	E	E		E	U	
Mineral Oil	F		E	E	G	G		E	E	E	E	G	G	E	U	U
Molasses (Food)	U	U	E		U	U		E		G		E			E	
Naptha	F	E	E	G	G	G		E	E	E	G	U	U	E	U	U
Napthalene	F	E	E	G	G	G	U	E		E	U	U	U	E	U	U
Natural Gas (Methane)	G	E	E		E	E	E	E			E	E	E	E	U	F
Nickel Ammonium Sulphate	U		E	F	U	U						E				
Nickel Chloride	U		G	G	U	U		G	E	E	E	E	E	E	E	E
Nickel Sulphate 10% 60°F	U		E	G	G	G	E	E	E	E	E	E	E	E	E	G
Nitric Acid 10% 70°F	U		G	U	U	U	U	G	G	U	U	G	E	E	G	U
Nitric Acid 30% 70°F	U		G	U	U	U	U	G	G	U	U	F	E	E	G	U
Nitric Acid 60% 175°F	U		G	U	U	U	U	G	G	U	U	U	U	E	U	U
Nitric Acid 70%	U	U	G	U	U	U	U	G	G	U	U	U	U	E	U	U
Nitrobenzene	U		E	G			E	E		F	U	U	U	G	U	
Oils and Fats	E	E	E	G	E	E		E			E			U	U	
Oils, Fish	U	G	E		F	F		E			G	E	F	U	U	
Oleic Acid 100°F	U	F	G	G	G	G	U	E	E	E	F	G	G	G		U
Oleum (Fuming Sulfuric Acid)	U	U	F					G	G		U	U	U	E	U	U
Oxalic Acid	U	F	G	F	F	F	G	G	G	F	G	G	G	E	G	G
Oxygen	E	E	E	E	E	E	E	E		E	G	E	E	E	E	
Ozone	U	G	E							E	U	F	E	E	E	U
Palmitic Acid	U	G	E	G	G	G		E		E	E	G	F	G	G	F
Perchlorethylene	F	G	E	G	G	G		E		E	F	U	U	E	U	U
Petroleum — Refined	G		E	E	E	E				E	E	G	E	F	U	U
Petroleum — Sour	U	G	G	G	F	F		G	E	E	F		E	G	U	
Phenol	U		E	F	G	G		E	E	E	U	U	U	G		U
Phosgene			E								G	U	U	G	U	U
Phosphoric Acid 10% 70°F	U	F	G	F	U	U	U	G	E	E	G	G	G	E	G	F
Phosphoric Acid 25% 70°F	U	U	G	F	U	U	U	G	G	E	F	G	G	E	E	U
Phosphoric Acid 75% 70°F	U	U	G	F	U	U	U		F	E	U	G	G	G	E	U
Phosphorous Oxychloride	U						E									
Pickling Sol. (20% Nitric-4HF)	U		G	G			G				U	U	U	G	F	U
Picric Acid 80% 70°F	U		G	U			G	E	E	E	G	E	E	G	E	G
Polyethylene Glycol Dimethyl Ether (Selexol)														U	G	U
Potassium Chloride	U		G	G	F	E		E		E	E	F	E	E	E	E
Potassium Cyanide	F	G	G	G	U	U		E		E	E	E	E	E	E	E
Potassium Hydroxide 5% 70°F	F	G	E	E	U	U		E	E	G	E	G	E	G	E	G
Potassium Nitrate 6% 68°F	F	G	E	G	F	F		E	E	E	E	E	G	E	E	E
Potassium Phosphate	U		G	G	F	F					E	E		E		
Potassium Sulphate 7% 180°F	F	G	E	E	G	E	E	E		E	E	E	G	E	E	
Potassium Sulfide	U	G	E	F	U	U		E		E	E	G	G	G	G	G
Potassium Sulfite	U		E					E			E	G	G	E	E	G
Propane	F	E	E	E	E	E		E		E	E	G	G	E	U	U
Resins and Rosins	U	E	E	G	E	E		E	E		F	F	F	E		F
SAE #10 Oil	G				E	E					E		E		U	
Sea Water 70°F	U	F	G	E	G	E		G	E	E	E	E	E	G	E	F
Sewage	U	G	E	G	G	G		E		E	E	E	E	E	G	F
Silicate Ester Synthetic Oil												G	E	E	U	
Skydrol 500	G	E	E		E	E		E			U	U	U	U	E	U
Soap Solution (Stearate) 70°F	U	F	E	E	E	E		E	E	E	E	E	E	E	E	F
Sodium Acetate 5% 75°F	U	F	E	G	G	G	E	E	E	E	G	G	U	U	E	E
Sodium Aluminate	U		E	G	G	G		E			E	E	G	E	G	G
Sodium Bisulfate	U	F	E					E		E	E	E	E	E	E	E
Sodium Carbonate 80% 60°F	U	G	E	E	G	G	U	E	E	E	E	E	E	E	E	E

Figure A-1 *(Continued)*

BUTTERFLY VALVE MATERIAL SELECTION GUIDE

APPLICATION CODE: **E** — Excellent
G — Good
F — Fair
U — Unsatisfactory

Stem & Disc Materials — Seat Materials

	Nickel Plated Ductile Iron	416SS	316SS	Monel	Manganeze Bronze	Aluminum Bronze	Lead	Alloy 20	Hastelloy C	PVF	Buna N	Neoprene	Hypalon	Viton	EPT	Natural Rubber
Sodium Chloride 30% 180°F	U	F	E	E	F	E	E	E	E	E	E	E	E	E	E	E
Sodium Cyanide	U		G	U	U	U	E	E		E	E	E	E	E	E	E
Sodium Fluoride 5% 60°F	U		G	G	F	F			E	E	E	E	G	E	G	G
Sodium Hydroxide 5%	U	G	E	G	F	F	U	E	E	G	G	F	E	G	E	E
Sodium Hydroxide 20%	U	F	E	E	F	F	U	E	E	G	G	U	E	G	E	E
Sodium Hydroxide 50%	U	F	G	G	U	U	U	E	E	E	G	U	G	F	G	G
Sodium Hydroxide 70%	U	U	F	F	U	U	U	G	G	G	F	U	U	U	G	F
Sodium Hypochlorite 5% 60°F	U		G	G	U	U	E	G	E	E	F	U	E	E	G	F
Sodium Nitrate 30% 60°F	U	G	E	G	G	G		E	E	E	G	G	E	E	E	G
Sodium Perborate	U	G	G	G	G	G					G	G	G	E	E	G
Sodium Peroxide	U	G	E	G	U	U				E	G	G	G	E	E	G
Sodium Phosphate 5% 60°F	U	G	E	G	F	F	E	E	E	E	E	G	E	E	E	E
Sodium Silicate	U	G	E	G	G	G		E	E	E	E	E	E	E	E	E
Sodium Sulphate 80% 60°F	U	G	E	G	G	G	U	E	E	E	E	E	E	E	E	G
Sodium Sulfide 70% 70°F	U	G	E		U		F	E	E	E	E	E	E	E	E	G
Sodium Sulfite 5% 70°F	U		E	G	U	U	E	E	E	E	E	E	E	E	E	G
Steam 225°F	U	U	E	E	E	E	U				U	U	U	U	G	U
Steam 300°F	U	U	E	E	G	G	U				U	U	U	U	U	U
Stearic Acid 90% 200°F	U	G	E	G	F	F	U	E	E	E	G	G	G	E	G	G
Steep Water													G	G		U
Styrene (Fumes)			E								U			G	U	
Sulfomic Acid								E		E			E		G	
Sulphur (Molten)	U	F	G	U	U	U	U	E	E		U	F	U	G	F	U
Sulphur Dioxide 60°F	U	G	E	E	F	F	U	E	E	E	U	G	F	E	E	G
Sulphur Trioxide	U	G	E				U	E	E		U	U	U	E	G	G
Sulfuric Acid 0-7% 70°F	U	F	G	F	U	U	E	E	E	E	U	G	E	E	U	U
Sulfuric Acid 7-40% 70°F	U	U	U	U	U	U	E	G	G	E	U	U	E	E	U	U
Sulfuric Acid 40-75% 70°F	U	U	U	U	U	U	E	G	G	E	U	U	G	G	U	U
Sulfuric Acid 75-95%	U	U	U	U	U	U	G	U	G	E	U	U	U	G	U	U
Sulfuric Acid 95-100%	U	U	U	U	U	U	G	U	E	E	U	U	G	G	U	U
Sulphurous Acid 80% 100°F	U	U	G	F	U	U	E	E	E	G	U	U	E	E	U	
Tall Oil	G	G	E	G				E		E	G	G		E		E
Tannic Acid 150°F	U	F	E	G	G	G		E	E	G	E	G	G	E	E	E
Tar	F	E	E	E	E	E	E	E	E		F	F		E	U	U
Tartaric Acid 150°F	U	G	E	G	F	G	E	E	E	G	G	G	E	E	G	E
Toluol and Toluolene	G	E	E		E	E		E		G	U	U	U	E	U	U
Toluene Sulfomic Acid			U							U				G		
Transformer Oil											U			E	U	
Tributyl Phosphate	G	F	E					E		E	U	U	U	U	E	G
Trichloroacetic Acid	U	U	U	G	U	G				F	G	U	U	F	G	F
Trichloroethylene	F	U	E	G	F	G		E		G	U	U	U	E	U	U
Triethinolamine	U	U	E	G						G	F	E	E	U	G	G
Trisodium Phosphate	U	U	G	G	F	F	E			E	F	F				
Tung Oil	U	U	E	F	E	E		E	E		E	G	G	E	U	U
Turpentine	U	U	E	G	G	G	F			E	U	F	F	E		
Vinegar 70°F	U		E	E				E	E	G	G	G	E	E	E	G
Water, Acid-Mine	U		E	U				E	E	G	F	F	E	E	F	F
Water – Deionized			E		E					E	E		E		E	
Water – Demineralized	U		E	E	E	E		E		E	E	E	E	E	E	F
Water – Fresh 180°F	G	E	E	E	E	E		E	E	E	G	G	F	U	E	U
Water – Fresh 225°F	G	E	E	E	E	E		E	E	E	U	U	U	U	E	U
Water - Salt 180°F	U	U	G	E	G	G		E	E	E	G	G	F	U	E	U
Water – Sewage 80°F	U	G	E	E	G	E		E	E		E			E	E	
Whiskey and Wines	U	U	E	E	G	G		E	E	E	E	E	G	E	E	
White Liquor	U		E					E	E		G		G			F
Xylene, Xylol	F	E	E		E	E		E		E	U	U	U	E	U	U
Zinc Chloride 5% 160°F	U	U	F	G	U	U	U	G	G	E	E	E	E	E	E	E
Zinc Phosphate			G							E	E				E	
Zinc Sulphate 25% 180°F	U	G	E	G	G	G	E	E	E	E	E	E	E	E	E	G

Figure A-1 *(Continued)*

The following application data is for use in determining the available material *most* likely to give satisfactory service and is not meant to imply any type warranty. However, certain factors such as temperature, concentration and combinations of chemicals may affect the fluid resistance of any material. In questionable applications, testing is always the most positive way to determine suitability.

Some materials are available in certain valve sizes and series only. Refer to the specific valve sections for availability.

Service Guide

Chemical	Carbon Steel	316 Stn.Stl.	Monel	Alloy 20	TFE
Barium Hydroxide	C	A	B	A	A
Barium Sulfate	C	B	B	B	A
Barium Sulfide	C	B	C	B	A
Beer	C	A	A	A	A
Beet Sugar Liquors	B	A	A	A	A
Benzaldehyde	C	B	B	B	A
Benzene	B	B	A	A	A
Benzoic Acid	D	B	B	B	A
Benzyl Chloride	B	B			A
Black Sulfate Liquor	C	B	B	B	A
Bleaching Powders	D	B	D	B	A
Blood		A	B	A	A
Borax	C	A	A	A	A
Bordeaux Mixture		A			A
Boric Acid	D	B	B	B	A
Brine	C	B	A	B	A
Bromine (Dry)	D	D	B	D	A
Bromine (Wet)	D	D	D	D	A
Bromic Acid					A
Bunker Oils (Fuel)	B	A	A	A	A
Butadiene	B	A	C	A	A
Butane	B	A	A	A	A
Buttermilk	D	A	D	A	A
Butyl Acetate	A	B	B	B	A
Butylene	A	A	A	A	A
Butyric Acid	D	B	B	B	A
Calcium Bisulfite	D	B	D	B	A
Calcium Carbonate	D	B	B	B	A
Calcium Chlorate	C	B	B	B	A
Calcium Chloride	C	B	B	B	A
Calcium Hydroxide	C	B	B	B	A
Calcium Hypochlorite	D	C	C	B	A
Calcium Nitrate		B			A
Calcium Phosphate		B			A
Calcium Sulfate	C	B	B	B	A
Calgon	see Sodium Metaphosphate				
Caliche Liquor	B				A
Camphor		B	C	C	A
Cane Sugar Liquors		A	B	A	A
Carbolic Acid (Phenol)	D	B	B	B	A
Carbonated Beverages	D	B	C		A
Carbon Bisulfide	B	B	B	B	A
Carbon Dioxide (Dry)	A	A	A	A	A
Carbonic Acid	D	B	B	B	A
Carbon Monoxide	A	A	A	A	A
Carbon Tetrachloride, dry	C	B	A	B	A
Carbon Tetrachloride, wet	D	B	B	B	A
Casein		B	C		A

Chemical	Carbon Steel	316 Stn.Stl.	Monel	Alloy 20	TFE
Castor Oil	B	A	A	A	A
Caustic Soda	see Sodium Hydroxide				
Caustic Potash	see Potassium Hydroxide				
Cellulose Acetate	B	B	B	B	A
China Wood Oil (Tung)	C	A	A	A	A
Chlorinated Solvents	C	B	B	B	A
Chlorinated Water		C	D		A
	B	B	A	B	A
Chlorine, Wet	D	D	C	D	A
Chloroacetic Acid	D	D	D	C	A
Chlorobenzene, dry	B	B	B	B	A
Chloroform,dry	B	A	A	A	A
Chlorophyll, dry		B	B		A
Chlorosulfonic Acid, dry	B	B	B	B	A
Chlorosulfonic Acid, wet	D	D	C	D	A
Chrome Alum	B	A	B	A	A
Chromic Acid	D	C	D	C	A
Chromic Chloride					
Chromium Sulfate		B	B	C	A
Citric Acid	D	A	B	A	A
Coconut Oil	C	B	B	B	A
Coffee	C	A	B	A	A
Coke Oven Gas	B	A	B	A	A
Cooking Oil	B	A	A		A
Copper Acetate	D	B	C	B	A
Copper Carbonate		A	A		A
Copper Chloride	D	D	C	D	A
Copper Cyanide		A	C		A
Copper Nitrate	D	B	D	B	A
Copper Sulfate	D	B	C	B	A
Corn Oil	C	B	B	B	A
Cottonseed Oil	C	B	B	B	A
Cresol		B		B	A
Creosote Oil	B	B	B	A	A
Cresylic Acid	C	B	B	B	A
Crude Oil, Sour	B	A	A	A	A
Crude Oil, Sweet	B	A	A	A	A
Cumene	B	B	B	B	A
Cupric Nitrate		A	D		A
Cutting Oils, Water Emulsions	B	A		A	A
Cyanide Solutions		B	D		A
Cyclohexane	A	A	A	A	A
Cyclohexanone	B	B	B	B	A
Denatured Alcohol	see Alcohol, Ethyl				
Detergents		B	B		A
Dextrin (Starch gum)		B	B		A
Diacetone Alcohol	A	A	A	A	A
Diamylamine		A		A	A

*Dry chlorine as defined by the Chlorine Institute is gaseous or liquid chlorine containing less than 150 PPM water.

A = Excellent B = Fair C = Poor D = Not Recommended

Figure A-2 Valve material and service guide. (*Courtesy of ITT Engineered Valves.*)

Service Guide

Chemical	Carbon Steel	316 Stn.Stl.	Monel	Alloy 20	TFE
Dibutyl Phthalate					A
Dichloroethane (Ethylene dichloride)	D	C	B	C	A
Dichloroethyl Ether		B			A
Diesel Fuels	A	A	A	A	A
Diethylamine	A	A	B	B	A
Diethylbenzene					A
Diethylene Glycol		A	B		A
Diethyl Sulfate		B	B		A
Dimethyl Formamide		A	B		A
Dioxane		B	B		A
Dipentane (Piñene)		A			A
Dowtherm	B	A	A	A	A
Drilling Mud	B	A	A	A	A
Dry Cleaning Fluids	B	A	B	A	A
Drying Oils	C	B	B	B	A
Enamel					A
Epichlorohydrin					
Epsom Salts	C	B	B	B	A
Esters, General		A			A
Ethane	C	B	B	B	A
Ethers	A	A	B	A	A
Ethyl Acetate	B	B	B	B	A
Ethyl Acrylate	C	A	B	A	A
Ethylamine		A			A
Ethyl Benzene	A	B	B	B	A
Ethyl Bromide		B	B	C	A
Ethyl Cellosolve					
Ethyl Chloride, dry	B	A	B	A	B
Ethyl Chloride, wet	D	B	B	B	A
Ethylene	A	A	A	A	A
Ethylenediamine		A			A
Ethylene Chloride	see Ethylene Dichloride				
Ethylene Dibromide	B	A	B	A	A
Ethylene Dichloride, Dry	B	A	A	A	A
Ethylene Dichloride, Wet	D	C	B	A	A
Ethylene Glycol	B	B	B	A	A
Ethylene Oxide	B	B	B	B	A
Ethyl Ether	A	A	B	A	A
Ethyl Silicate		B	B		A
Ethyl Sulfate					A
Fatty Acids	D	A	B	A	A
Ferric Chloride	D	D	D	D	A
Ferric Hydroxide		A	A		A
Ferric Nitrate	D	C	D	A	A
Ferric Sulfate	D	B	D	A	A
Ferrous Chloride	D	D	D	D	A
Ferrous Sulfate	D	B	C	B	A
Fertilizer Solutions	B	B	B	B	A
Fish Oils	B	A	A	A	A
Flotation Solutions					A
Flue Gases		A	B		A
Fluoboric Acid		B	B		A
Fluoride Salts		B			A
Fluorine, dry	B	B	B	B	A
Fluosilicic Acid	D	B	A	B	A
Food Fluids & Pastes	C	A	A	A	A
Formaldehyde, cold	A	A	A	A	A
Formaldehyde, hot	D	C	B	A	A
Formic Acid, cold	D	B	B	A	A
Formic Acid, hot	D	B	B	B	A
Freon, dry	B	A	A	A	A
Freon, wet	D	C	B		A
Fruit Juices	D	A	B	A	A
Fuel Oil	B	A	B	A	A
Furfural	A	A	B	A	A
Gallic Acid	D	B	B	B	A
Gas, Manufactured	B	B	A	B	A
Gas, Natural	B	A	A	A	A
Gas, Odorizers	B	B	B	B	A
Gasoline, Aviation	A	A	A		A
Gasoline, Leaded	A	A	B		A
Gasoline, Refined	B	A	B	A	A
Gasoline, Sour	B	A	C	A	A
Gasoline, Unleaded	A	A	A		A
Gelatin	D	A	B	A	A
Glucose	B	A	A	A	A
Glue	A	B	A	A	A
Glutamic Acid		B		A	A
Glycerol (Glycerin)	B	A	A	A	A
Glycol Amine		B			A
Glycol	B	B	B	A	A
Grease	A	A	B	A	A
Helium Gas		B	B		A
Heptane	B	A	B	B	A
Hexane	B	B	B	B	A
Hexanol, Tertiary	A	A	A	A	A
Hydraulic Oil, Petroleum Base	A	A	A	A	A
Hydrazine		B	D		A
Hydrobromic Acid	D	D	D	D	A
Hydrochloric Acid	D	D	D	D	A
Hydrocyanic Acid	D	B	C	A	A
Hydrofluoric Acid	D	D	B	D	A
Hydrofluosilicic Acid	D	C	B	B	A
Hydrogen Gas, Cold	B	A	A	A	A
Hydrogen Gas, Hot	B	B		A	A
Hydrogen Bromide Gas					

A = Excellent B = Fair C = Poor D = Not Recommended

Figure A-2 *(Continued)*

Service Guide

Chemical	Carbon Steel	316 Stn.Stl.	Monel	Alloy 20	TFE
Hydrogen Chloride Gas(Dry)	D	B	B	A	A
Hydrogen Peroxide	D	B	C	B	A
Hydrogen Sulfide, Dry	B	A	B	B	A
Hydrogen Sulfide, Wet	C	B	C	B	A
Hypo (Sodium Thiosulfate)	D	A	B	B	A
Hypochlorous Acid					A
Illuminating Gas	A	A	A		A
Ink	D	A	B	A	A
Iodine	D	D	D	D	A
Iodoform	B	A	C	A	A
Isobutane		B			A
Isooctane	A	A	A	A	A
Isopropyl Acetate	B	B	B	B	A
Isopropylamine		B			A
Isopropyl Ether	A	A	B	A	A
Jet Fuels (JP-4,-5,-6)	A	A	A	A	A
Juice, Fruit	see Fruit Juice				
Juice, Citric	see Citric Acid				
Kerosene	B	A	A	A	A
Ketchup	D	A	B	A	A
Ketones	A	A	A	A	A
Lacquers (& Solvents)	C	A	A	A	A
Lactic Acid (Conc, Cold)	D	A	D	A	A
Lactic Acid (Conc Hot)	D	B	D	A	A
Lactic Acid (Dil, Cold)	D	A	C	A	A
Lactic Acid (Dil, Hot)	D	A	D	A	A
Lactose		B	B		A
Lard		A			A
Lard Oil	C	B	B	A	A
Latex		A			A
Lead Acetate	D	B	B	B	A
Lead Arsenate		B			
Lead Oxide					
Lead Sulfate		B	B		A
Lecithin		B	B		A
Linoleic Acid	B	A	B	A	A
Linseed Oil	A	A	B	A	A
Liquified Petroleum Gas	B	B	B	B	A
Lithium Chloride		B	B	A	A
Lubricating Oil(PetroleumBase)	A	A	B	A	A
Ludox		B	B		A
Lye	see Sodium Hydroxide & Potassium Hydroxide				
Magnesium Bisulfate	B	A	B	A	A
Magnesium Bisulfide		B	B		A
Magnesium Carbonate		A	B		A
Magnesium Chloride	C	B	B	B	A
Magnesium Hydroxide	B	A	A	A	A
Magnesium Nitrate		A	B		A
Magnesium Oxide					

Chemical	Carbon Steel	316 Stn.Stl.	Monel	Alloy 20	TFE
Magnesium Sulfate	B	A	B	A	A
Maleic Acid	B	B	B	B	A
Maleic Anhydride	B	B	B	B	A
Malic Acid	D	B	B	B	A
Manganese Carbonate		B			A
Manganese Sulfate		B	B		A
Mayonnaise	D	A	B	A	A
Melamine Resins		C			A
Menthol		B	B		A
Mercaptans	B	B			A
Mercuric Chloride	D	C	D	C	A
Mercuric Cyanide	D	B	C	B	A
Mercuric Nitrate		A	D	A	A
Mercury	A	A	B	A	A
Methane	B	B	B	A	A
Methanol	see Alcohol Methyl				
Methyl Acetate	B	A	B	B	A
Methyl Acetone	A	A	A	A	A
Methylamine	B	A	C	B	A
Methyl Bromide		B	B		A
Methyl Cellosolve	B	A	B	B	A
Methyl Chloride	B	A	B	A	A
Methylene Chloride	B	A	B	A	A
Methyl Ethyl Ketone	A	A	A	A	A
Methyl Formate	C	B	B	B	A
Methyl Isobutyl Ketone		B			A
Milk	D	A	B	A	A
Mineral Oils	B	A	A	A	A
Mineral Spirits	see Naptha				
Mine Water (Acid)	D	C	C	C	A
Mixed Acids (Nitric-Sulfuric)	C	B	C	B	A
Molasses, Crude	A	A	A	A	A
Molasses, Edible	C	A	A		A
Molybdic Acid		A			A
Monochloroacetic Acid	see Chloroacetic Acid				
Monochlorobenzene	see Chlorobenzene				
Morpholine		B	B		A
Muriatic Acid	see Hydrochloric Acid				
Mustard	B	A	A	A	A
Naptha	B	B	B	B	A
Napthalene	B	B	B	B	A
Natural Gas	B	A	D	A	A
Nickel Ammonium Sulfate	D	A	C	A	A
Nickel Chloride	D	B	B	B	A
Nickel Nitrate	D	B	B	A	A
Nickel Sulfate	D	B	B	B	A
Nicotinic Acid	B	A	A		A
Nitric Acid 10%	D	A	D	A	A
Nitric Acid 30%	D	A	D	A	A

A = Excellent B = Fair C = Poor D = Not Recommended

Figure A-2 *(Continued)*

Service Guide

Chemical	Carbon Steel	316 Stn.Std.	Monel	Alloy 20	TFE
Nitric Acid 80%	D	B	D	C	A
Nitric Acid 100%	D	A	D	A	A
Nitric Acid, Anhydrous	D	A	D	B	A
Nitrobenzene	B	A	B	A	A
Nitrocellulose	B	B	B	B	A
Nitrogen Gas	A	A	A	A	A
Nitrous Acid (10%)	D	B	D	B	A
Nitrous Gases	B	A	D	A	A
Nitrous Oxide	B	B	D	B	A
Oils, Animal	A	A	A	A	A
Oils, Deisel	B	A	A	A	A
Oils, Edible		A		A	A
Oil, Emulsified					A
Oil, Fish	B	A	A	A	A
Oil, Fuel	B	A	B	A	A
Oil, Linseed	A	A	B	A	A
Oil, Lubricating	A	A	B	A	A
Oil, Petroleum Refined	A	A	A	A	A
Oil, Petroleum, Sour	B	A	A	A	A
Oleic Acid	C	B	B	B	A
Oleum	see Sulfuric Acid, Fuming				
Olive Oil	B	A	A	A	A
Oxalic Acid	D	B	B	B	A
Oxygen	B	A	A	A	A
Ozone, Dry	A	A	A	A	A
Ozone, Wet	C	A	A	A	A
Paints & Solvents	A	A	A	A	A
Palmitic Acid	C	B	B	B	A
Palm Oil	C	B	A	A	A
Paper Pulp		B	B		A
Paraffin	B	A	A	A	A
Paraformaldehyde	B	B	B	B	A
Paraldehyde					A
Pentane	B	A	B	B	A
Perchloroethylene, Dry	B	B	A	B	A
Petrolatum	C	B	A	B	A
Petroleum, Refined	A	A	A	A	A
Petroleum, Sour	B	A	A	A	A
Phenol	B	A	A	A	A
Phosphoric Acid (10% cold)	D	C	C	B	A
Phosphoric Acid (10% hot)	D	C	B	B	A
Phosphoric Acid (50% cold)	D	B	C	B	A
Phosphoric Acid (50% hot)	D	D	C	B	A
Phosphoric Acid (85% cold)	D	A	A	B	A
Phosphoric Acid (85% hot)	C	B	C	B	A
Phosphoric Anhydride		A			A
Phosphorous Oxychloride		B			A
Phosphorous Trichloride (dry)	B	A			A
Phthalic Acid	C	B	B	B	A

Chemical	Carbon Steel	316 Stn.Std.	Monel	Alloy 20	TFE
Phthalic Anhydride	C	B	A	B	A
Pickling Solutions	refer to Specific Acids				
Picric Acid	D	B	D	B	A
Pineapple Juice	C	A	A	A	A
Pine Oil	B	A	A	A	A
Polysulfide Liquor	C	B	B		A
Polyvinyl Acetate		B	B		A
Polyvinyl Chloride		B	B		A
Potassium Bicarbonate	B	A	B		A
Potassium Bichromate	see Potassium Dichromate				
Potassium Bisulfate		A	B		A
Potassium Bisulfite	D	B	D	B	A
Potassium Bromide	D	B	B	B	A
Potassium Carbonate	B	B	B	B	A
Potassium Chlorate	B	B	C	B	A
Potassium Chloride	C	B	B	A	A
Potassium Chromate		B	B	B	A
Potassium Cyanide	B	B	B	B	A
Potassium Dichromate	C	B	B	A	A
Potassium Diphosphate	A	A	B	A	A
Potassium Ferricyanide	C	B	B	B	A
Potassium Ferrocyanide	C	B	B	B	A
Potassium Hydroxide (dilute, cold)	B	B	A	B	A
Potassium Hydroxide (to 70%, cold)	B	B	A	B	A
Potassium Hydroxide (dilute, hot)	B	B	A	B	A
Potassium Hydroxide (to 70%, hot)	B	B	A	B	A
Potassium Iodide	C	B	C	B	A
Potassium Nitrate	B	B	B	B	A
Potassium Oxalate	D	A		A	A
Potassium Permanganate	B	B	B	B	A
Potassium Phosphate (Monobasic)	B	B	B	B	A
Potassium Phosphate (Di-Basic)	A	A	B		A
Potassium Phosphate (Tri-Basic)	A	B	B	B	A
Potassium Sulfate	B	A	B	A	A
Potassium Sulfide	C	B	D	B	A
Potassium Sulfite	C	B	D	B	A
Producer Gas	B	B	A	B	A
Propane Gas	A	A	A	A	A
Propionic Acid	D	B	B	B	A
Propyl Alcohol	B	A	A	A	A
Propyl Bromide		B	B	A	A
Propylene Glycol	B	B	B	B	A

A = Excellent B = Fair C = Poor D = Not Recommended

Figure A-2 *(Continued)*

Service Guide

Chemical	Carbon Steel	316 Stn.Stl.	Monel	Alloy 20	TFE
Pyridine		B		A	A
Pyrogallic Acid	B	B	B	B	A
Pyroligneous Acid		A		B	A
Quench Oil	B	A	A		A
Quinine Bisulfate, dry	D	A	B		A
Resins & Rosins	C	B	A	B	A
Resorcinol		B	B		A
Road Tar	A	A	A	A	A
Roof Pitch	A	A	A	A	A
Rosin Emulsion	C	A	A		A
RD – 1 Fuel	A	A	A		A
Rubber Latex Emulsions	B	A		A	A
Rubber Solvents	A	A	A	A	A
Salad Oils	C	B	B	B	A
Salicylic Acid	D	A	B	A	A
Salt	C	B	A	B	A
Salt Brine		A	B		A
Sea Water	D	B	A	B	A
Sewage	C	B	B		A
Shellac - Bleached	A	A	A	A	A
Shellac - Orange	A	A	A	A	A
Silicone fluids		B			A
Silver Bromide	D	A	B		A
Silver Cyanide		A	B		A
Silver Nitrate	D	A	D	A	A
Soap Solutions (Stearates)	A	A	A	A	A
Sodium Acetate	C	B	C	B	A
Sodium Aluminate	C	A	A	B	A
Sodium Benzoate			B		A
Sodium Bicarbonate	C	D	B	A	A
Sodium Bichromate		A		B	A
Sodium Bisulfate (10%)	D	A	B	A	A
Sodium Bisulfite (10%)	D	A	B	B	A
Sodium Borate	C	B	B	B	A
Sodium Bromide (10%)	C	B	D	B	A
Sodium Carbonate (Soda Ash)	B	B	B	A	A
Sodium Chlorate	C	D	B	B	A
Sodium Chloride	C	B	A	B	A
Sodium Chromate	B	A	B	B	A
Sodium Citrate		A		B	A
Sodium Cyanide	B	B	B	A	A
Sodium Ferricyanide		A	B	B	A
Sodium Fluoride	D	B	B	B	A
Sodium Hydroxide (Cold, 20%)	A	A	A	A	A
Sodium Hydroxide (Hot, 20%)	B	A	A	A	A

Chemical	Carbon Steel	316 Stn.Stl.	Monel	Alloy 20	TFE
Sodium Hydroxide (Cold, 50%)	A	A	A	A	A
Sodium Hydroxide (Hot, 50%)	B	B	A	A	A
Sodium Hydroxide (Cold, 70%)	B	A	A	B	A
Sodium Hydroxide (Hot, 70%)	D	B	A	B	A
Sodium Hypochlorite	D	C	D	C	A
Sodium Hyposulfite	D	B	B		A
Sodium Lactate		A	B		A
Sodium Metaphosphate	B	B	B	B	A
Sodium Metasilicate	D	A	A	A	A
Sodium Nitrate	B	A	B	A	A
Sodium Nitrite	B	B	C	B	A
Sodium Perborate	B	B	B	B	A
Sodium Perchlorate		B		B	A
Sodium Peroxide	C	B	B	B	A
Sodium Persulfate		B			A
Sodium Phosphate (Monobasic)	B	B	B	B	A
Sodium Phosphate (Dibasic)	B	B	B	B	A
Sodium Phosphate (Tribasic)	B	B	B	B	A
Sodium Polyphosphate		B			A
Sodium Salicylate		A			A
Sodium Silicate	B	B	B	B	A
Sodium Stannate		B			A
Sodium Stearate		B			A
Sodium Sulfate	B	A	B	A	A
Sodium Sulfide	B	B	B	B	A
Sodium Sulfite	B	A	B	A	A
Sodium Thiosulfate	C	B	B	B	A
Sorbic Acid		D			A
Soybean Oil	C	A	A	A	A
Stannic Chloride	D	D	D	D	A
Stannous Chloride	D	B	C	A	A
Stannour Fluoride		B		B	A
Starch Solutions	C	B	A	A	A
Steam (212°F)	A	A	B		A
Stearic Acid	C	B	B	B	A
Styrene	B	A	B	A	A
Succinic Acid	B	B		B	A
Sugar Liquids	B	A	A	A	A
Sulfate Black Liquor	C	B	B	B	A
Sulfate Green Liquor	C	B	B	B	A
Sulfate White Liquor	C	B	C	B	A
Sulfite Liquors	D	B	D		A

A = Excellent B = Fair C = Poor D = Not Recommended

Figure A-2 *(Continued)*

Service Guide

Chemical	Carbon Steel	316 Stn.Stl.	Monel	Alloy 20	TFE
Sulfonic Acid		B	B		A
Sulfur	C	B	B	A	A
Sulfur Chlorides	D	D	B	B	A
Sulfur Dioxide (dry)	B	A	B	B	A
Sulfur Dioxide (wet)	D	A	C	B	A
Sulfur Hexafluoride		A			A
Sulfur Trioxide (dry)	B	B	B	B	A
Sulfuric Acid (0 - 7%)	D	B	C	C	A
Sulfuric Acid (7 - 40%)	D	D	C	B	A
Sulfuric Acid (40 - 75%)	D	D	C	B	A
Sulfuric Acid (75 - 95%)	C	D	D	B	A
Sulfuric Acid (95 - 100%)	C	C	D	B	A
Sulfurous Acid	D	B	D	B	A
Synthesis Gas	B	B	A	B	A
Tall Oil	B	B	B	B	A
Tallow, Molten	B	A		A	A
Tannic Acid	C	B	B	B	A
Tanning Liquors		B		B	A
Tar - Tar Oils	A	A	A		A
Tartaric Acid	D	A	B	B	A
Tetraethyl Lead	C	B	A	B	A
Thioglycolic Acid		B		B	A
Toluene (Toluol)	A	A	A	A	A
Toluene Diisocyanate		A		A	A
Tomato Juice	C	A	A	A	A
Transformer Oil	A	A	A	A	A
Tributyl Phosphate	A	A	A	A	A
Trichloroethane (Dry)	B	B	B	B	A
Trichloroethylene	B	B	A	B	A
Trichloroacetic Acid	D	D	C	B	A

Chemical	Carbon Steel	316 Stn.Stl.	Monel	Alloy 20	TFE
Triethanolamine	C		B	B	A
Triethylamine					A
Triethylene Glycol	C	B	B	B	A
Tri-Sodium Phosphate	B	B	B	B	A
Tung Oil	B	A	C	A	A
Turpentine	B	B	B	B	A
Urea	C	B	B	B	A
Uric Acid	D	A		B	A
Varnish	C	A	A	A	A
Vegetable Oils	B	A	B	A	A
Vinegar	D	A	B	A	A
Vinyl Acetate	A	B	B	B	A
Vinyl Chloride (dry)	A	B	B	B	A
Water, Acid Mine	D	B	D	A	A
Water, Deionized	D	A	A	A	A
Water, Distilled	D	A	A	A	A
Water, Fresh	C	A	A	A	A
Water,Sea	D	B	A	B	A
Wax Emulsions	A	A	A	A	A
Waxes	A	A	A	A	A
Whiskey & Wines	D	A	B	A	A
White Water	see Sulfate, White Liquor				
Xylene (Xylol) Dry	B	A	A	A	A
Zinc Acetate		B			A
Zinc Bromide		B	B		A
Zinc Chloride	D	D	B	A	A
Zinc Hydrosulfite	A	A	B	A	A
Zinc Sulfate	D	B	B	A	A

Figure A-2 *(Continued)*

Appendix

B

System Conversions

TABLE B.1 Unit Conversions

Length
1 in = 25.4 mm (exact)
1 mm = 0.03937 in
Pressure
1 bar = 14.5 psi
1 psi = 0.06895 bar
1 bar = 100 kPa
Force
1 lb (force) = 4.448 N
1 N = 0.2248 lb (force)
Torque
1 lb · ft = 1.356 N · m
1 N · m = 0.7376 lb · ft
Temperature
$°C = \frac{°F-32}{1.8}$
°F = 1.8°C + 32

TABLE B.2 Pressure and Class Number Identification

Customary Class	Metric PN	Reference (Source Document)	Notes†
	10	ISO-2084	
	16	ISO-284	
150	20	ANSI-Bl6.5 (Cl 150)	See Note 1
300	50	ANSI-Bl6.5 (Cl 300)	See Note 2
400	68	ANSI-Bl6.5 (Cl 400)	See Note 3
600	100	ANSI-Bl6.5 (Cl 600)	See Note 4
900	150	ANSI-Bl6.5 (Cl 900)	
1500	250	ANSI-Bl6.5 (Cl 1500)	
2500	420	ANSI-Bl6.5 (Cl 2600)	

†Notes

Note 1. Mating dimensions are similar to ANSI Bl6.1, Cl 25 and Cl 125.

Note 2. Mating dimensions are similar to ANSI Bl6.1, Cl 250.

Note 3. PN 68 (ANSI Bl6.5, Cl 400) is not recognized in ISO standards but will continue in use in USA standards.

Note 4. Mating dimensions are similar to ANSI Bl6.1, Cl 800.

Glossary

ac Alternating current or voltage which changes direction and polarity rhythmically. The most common alternating current used in industry is sinusoidal.

Accumulator tank Enclosed volume that contains a compressible fluid (gas) to provide standby energy in the event of loss of system pressure.

Air manifold Enclosure containing inlet and outlet ports that route the flow of air to the active piston or diaphragm areas.

Air pressure Available air supply at the user site; may be less than compressor setting.

Ampere Unit of measurement of electron flow or current. The rating on the motor nameplate gives the amount of current the motor draws from the power lines while operating at full load at rated voltage.

Analog Measurement system where the units and signals bear a direct linear relationship to functions.

Automatic system System which will function according to predetermined requirements without the intervention of human surveillance.

Auto transformer Static transformer having a single winding, part of which is common to both primary and secondary circuits.

Auxiliary switches Auxiliary circuit switches fitted and mechanically coupled to switching devices for the purpose of indicating the state of the main contacts.

Ball, V-notch Flow-controlling member for a popular style of throttling ball valve. The V-notch ball includes a polished or plated partial-sphere surface that rotates against the seal ring.

Ball valve Device that incorporates a ball with a center hole to provide shutoff (center hole perpendicular to pipe) or flow (center hole in line with pipe) through a pipe system. One of several types of 90° operation valves.

Bimetal strip Used in overload relays and overcurrent releases. It is a strip made up of two kinds of metal with different temperature coefficients. It is enclosed by a current winding. The heat due to an overload causes the strip to deflect and to actuate a tripping bar which, in turn, acts on the auxiliary circuit switch of the release or relay.

Black box Unit that converts input signals to appropriate output values without the need for the observer to know the internal workings of the device.

Breakaway (unseating) torque Published value depicting the turning force required to initiate rotation of the closure element of a rotary valve.

Butterfly valve Device that incorporates a circular disc to provide shutoff or modulating flow through a pipe system. One of several types of 90° operation valves.

Capacitance The property of a capacitor device to store electric charge. The unit of measurement is the farad-coulomb per volt = ampere per second per volt.

Capacitor Device having two conductive plates separated by a dielectric insulation substance and thus possessing the property of capacitance.

Capacity Rate of flow through a valve under stated conditions.

Cavitate Formation and subsequent collapse of voids or cavities in a valve resulting from increased fluid velocity through the restricted areas of the valve. It occurs in liquids when the valve operates near the vapor pressure of the liquid.

Characteristic Relation between flow through the valve and percent rated travel as the latter is varied from 0 to 100 percent.

Circuit Continuous path for an electric current to flow through.

Circuit breaker Switch with a making and breaking capacity, capable of controlling items of equipment in normal operation and under fault conditions.

Closure member Moveable part of the valve which is positioned in the flow path to modify the rate of flow through the valve.

Conductance Reciprocal of resistance. Measured in siemens.

Conductor Material in which free electrons are available to move under the influence of an electric field.

Contact Condition brought about by two conductive parts (e.g., contact pieces) touching. The expression is also used to denote a contact element or contact piece.

Contactor Remote-control switch actuated electromagnetically and capable of switching motor loads.

Control circuit Auxiliary circuit.

Control switches Power circuit switches for the direct control of electrical equipment, or auxiliary circuit switches for indirect control.

Control valve Power-operated device which modulates the fluid flow rate in a process control system. It consists of a valve connected to an actuator mechanism that is capable of changing the position of a flow-controlling element in the valve in response to a signal from the controlling system.

Corrosion Damaging effects of hostile media on control valve components resulting from material incompatibility.

Current Flow of electrons through an electric circuit.

C_g Flow coefficient commonly used for gases.

C_s Flow coefficient commonly used for steam.

C_v Flow coefficient commonly used for liquids. See *flow coefficient.*

dc Direct current or voltage which has a constant value related to zero value. Can be stipulated as a positive or negative value.

Deadband Range through which input can be varied without initiating observable response.

Diagram

Block Diagrammatic representation of a circuit or system in which the various component parts of the system are represented by a rectangle or block with no detail shown for the complexities contained in the system block.

Ladder or schematic Diagrammatic representation of a circuit or system where for the sake of clarity no relationship is made to actual physical location of components.

Wiring Diagrammatic representation of a circuit or system where actual physical location of components with regard to each other is shown.

Diaphragm Flexible seal area that provides for linear motion by the change of shape of the material.

Digital System that handles information as numbers.

Disc Flow-controlling member used in the most common varieties of butterfly rotary valves.

Disc, eccentric Commonly referred to as a high-performance butterfly valve. See *high performance butterfly valve.*

Double acting Characteristic of a piston or diaphragm system in which the energizing pressure acts on both faces of the piston or diaphragm and operates the system in forward and reverse to open or close.

Duty cycle Percentage of time a device is allowed to operate over a given period of time. Expressed in percent, it equals time *on* divided by time *off* multiplied by 100.

Efficiency Ratio of useful output power to total input power.

Electromagnet Device in which a core of ferromagnetic material is encircled by a coil of wire carrying an electric current. The core can be made magnetic only when current is passing through the coil.

Electromotive force Force which causes the movement of electric charges. Measured in volts.

Electronics That part of electrical engineering which specializes in the conductance properties of semiconductor materials.

Erosion Damaging effects of gritty or dirty fluids on control valve components. Erosion is forestalled with valve designs which separate flow stream from critical valve components and with hardened materials.

Explosion-proof Characteristic of a device or element that inherently contains or prevents an explosion.

Farad Unit of capacitance.

Fail safe Characteristic of a system that provides that a unit will be operated into a safe condition upon the failure of electric power, power source, etc.

Failure mode Upon electric power failure or air pressure loss to the actuator, the valve is operated to a predetermined position (fail open, fail closed, fail in last position).

Feedback Signal indicating the actual position or the electrical or pneumatic condition of an element in a control system.

Flangeless body Body style common to rotary-shaft control valves. Flangeless bodies are held between ANSI class flanges by through bolts. (Sometimes also called wafer-style valve bodies.)

Flashing Phenomenon observed in liquid service when the pressure of the medium falls below its vapor pressure and does not recover to a higher pressure.

Flow characteristic Relationship between flow through the valve and percent rated travel as the latter is varied from 0 to 100 percent. This is a special term. It should always be designated as either inherent flow characteristic or installed flow characteristic. Common flow characteristics are linear, equal percentage, and quick opening. See *Inherent flow characteristic* and *installed flow characteristic.*

Flow coefficient C_v The number of U.S. gallons per minute of 60°F water that will flow through a valve with a 1 lb/in^2 pressure drop.

Four-way pilot valve Pilot valve used with double-acting actuators.

Fusible element A specially shaped copper conductor (silver is used in some cases) in the ceramic body of the fuse element. It melts on overload or short circuit. Its special features are the low-melting-point insert and the so-called necks. The insert melts on overload, and the necks will interrupt a short circuit.

Globe valve Valve construction style with a linear motion flow-controlling member with one or more ports, normally distinguished by a globe-shaped cavity around the port region. Two categories are commonly recognized depending on the method of plug guiding: cage-guided and stem- or plug-guided.

Ground Main body of the earth which is considered to be at a zero electric potential. The ground itself and any conductor connected to it are termed at zero potential.

Hard facing Process of applying a material harder than the surface to which it is applied. This technique is used to resist fluid erosion and/or to reduce the chance of galling between moving parts, particularly at high temperature.

Hazardous environment Any combination of combustible or potentially explosive mixtures existing in the surrounding atmosphere.

Hertz Unit of frequency. Number of times per second that alternating current flows in one direction, reverses, and flows in the other direction. Power companies in the United States hold this figure constant at 60 cycles per second (hertz).

High-performance butterfly valve (HPBV) Common name for valve design in which the positioning of the valve shaft-disc connections causes the disc to take a slightly eccentric path on opening. (This allows the disc to be swung out of contact with the seal as soon as it is opened, thereby reducing friction and wear.) See *disc, eccentric.*

Horsepower Unit of power equal to 746 W, or 33,000 ft · lb of work per minute. The relationship between horsepower and torque is horsepower equals torque times revolutions per minute divided by 5250.

Hydrodynamic torque Turning force exerted on a valve stem due to fluid flow through the valve and its effect on the closure element.

Inductance Property of a circuit to generate an electromagnetic force by reason of a change in the magnetic flux through the circuit.

Inherent flow characteristic Flow characteristic when constant pressure drop is maintained across the valve.

Inlet Body opening through which fluid enters the valve.

In-rush current 1. Current input of a motor on starting. 2. Current input of a solenoid or contactor.

Installed flow characteristic Flow characteristic when pressure drop across the valve varies as dictated by flow and related conditions in the system in which the valve is installed.

Insulation Nonconductive material, e.g., vacuum, glass, nylon, rubber.

Integrated circuit Semiconductor device containing circuit elements which are manufactured in a single piece of material and which are individably connected to perform a function.

Leakage Quantity of fluid passing through an assembled valve when the valve is in the fully closed position under stated closure forces, with pressure differential and pressure as specified. Leakage is usually expressed as a percentage of the valve capacity at full rated travel.

Limit switch Set of electric contacts that are activated mechanically at a preselected position.

Linear flow characteristic Inherent flow characteristic which can be represented ideally by a straight line on a rectangular plot of flow coefficient versus percent rated travel. (Equal increments of travel yield equal increments of flow coefficient change at a constant pressure drop.)

Line voltage Voltage existing between the two lines of an electric supply.

Manual override Capability provided to actuators that allows the manual positioning of the actuator.

Modulate Function of a controller which causes a valve to respond to an infinite number of positions between the closed and full-open positions.

Moment arm Linear distance from the center of rotation upon which a force may act to produce torque.

Multiple-spring design Unique design where two or more springs may be used in parallel to provide a variable spring rate within a fixed space.

NEMA classification Code established for the construction of electrical components by the National Electrical Manufacturers Association.

Noise, control valve Generally refers to aerodynamic noise associated with flow-stream turbulence in compressible fluids. Noise levels can be reduced to safe levels defined by OSHA and the EPA with noise-abatement trim (source treatment) and with silencers and diffusers (path treatment).

Normally closed control valve One which closes when the actuator pressure is reduced to atmospheric.

Ohm Measure of resistance. Even in a good conductor, such as copper wire, there is some resistance to current flow. Three factors determine this resistance: the conductor's substance, its diameter, and its length. Large-diameter wiring offers less resistance than smaller-diameter wiring, and the farther the current has to flow from the power source to the equipment, the greater the resistance and the greater the voltage drop.

Ohm's law Explains the relationship of amperes, volts, and ohms. One ampere of electric current will flow when pushed by 1 V against 1 Ω of resistance. Amperes = volts/ohms.

On-off valve Any of a number of valve types used for either full-open or shutoff service. Terminology indicates an automated assembly.

Outlet Body opening through which fluid exits the valve.

Output torque characteristics Equation or graph that shows torque versus position, usually 0 through 90° operation.

Overload Conditions in a healthy electric circuit that result in an overload of current flow. A sustained overload may cause the circuit to suffer damage.

Overload relay Relay which operates as a function of the current. Used for the protection of electrical equipment (e.g., motors) against overload. Its mode of operation is to actuate a built-in auxiliary switch by thermal or magnetic means. The auxiliary switch causes the switching device to be tripped remotely. Overload relays must be protected by fuses to prevent them from being destroyed by a short circuit.

Packing box (assembly) Cavity in the valve body or the part of the bonnet assembly used to seal against leakage around the valve plug stem or shaft.

Parallel Two or more circuits which are said to be connected in parallel when the total current flow is divided between them.

Pilot valve Device acting between the source of air pressure and the actuator that directs air flow to the required actuator air inlet ports.

Piston area Net cross-sectional area upon which air pressure may act to provide an output force.

Plug, eccentric Flow-controlling member of the eccentric rotary plug valve. Because of its eccentric action, it clears its seat soon after opening.

Plug valve Device that uses a truncated V-shaped element to provide tight shutoff or flow through a pipe system. One of several types of 90° operation valves.

Port Fixed opening, normally the inside diameter of a seat ring, through which fluid passes.

Power supply There are two types of power supply: direct current and alternating current. The latter is the most widely used power supply.

Pressure drop Difference between upstream pressure and downstream pressure that represents the amount of flow-stream energy that the control valve must be able to withstand.

Pressure drop, maximum allowable Maximum flowing or shutoff pressure drop that a control valve can withstand. While maximum inlet pressure is commonly dictated by the valve body, maximum allowable pressure drop is generally limited by the internal controlling components (plug, stem, disc, shaft, bearings, seals). Maximum allowable pressure drop may apply to the pressure drop while flowing process fluids or at shutoff.

Process control Act of maintaining or correcting a process within preset limits using a measurement of the actual process conditions in comparison to an established, desired condition.

Proportional controller Device that provides an output signal based on a valve(s) or a measured process parameter(s) and an input set-point command.

Quick-opening flow characteristic Inherent flow characteristic in which there is maximum flow with minimum travel.

Rack Linear gear that operates in conjunction with a pinion gear to provide rotary motion.

Rangeability Ratio of maximum to minimum flow within which the deviation from the specified inherent flow characteristic does not exceed some stated limit. Rangeability is expressed as the ratio of the maximum flow coefficient to the minimum usable flow coefficient of a control valve.

Rated C_v Value of C_v at the rated full-open position.

Rating Specified operating conditions or performance of a piece of equipment by way of its input-output values.

Recovery Relative term used to describe how much flow-stream pressure is reduced because of the design of the control valve; the ratio of maximum (valve fully open) downstream pressure to upstream pressure.

High-recovery valve: A valve design that dissipates relatively little flow-stream energy due to streamlined internal contours and minimal flow turbulence. Therefore, pressure downstream of the valve vena contracta recovers to a pressure significantly higher than the pressure at the vena contracta. (Straight-through flow valves, such as rotary-shaft ball and butterfly valves, are typically high-recovery valves.)

Low-recovery valve: A valve design that dissipates a considerable amount of flow-stream energy due to turbulence created by the contours of the flow path. Consequently, downstream pressure recovery is reduced when compared to high-recovery valves. (Although individual designs vary, conventional globe-style valves generally have low-pressure recovery capability.)

Relays May be integral parts of, or used as accessory items for, individual switching devices. They operate as a function of the current or voltage or of time to control other circuit elements connected through auxiliary circuits.

Resistance Opposition to the flow of electric current exhibited by matter.

Rotary actuator Device that provides torque or rotating movement to a rotary valve.

Rotary-shaft control valve Valve style in which the flow closure member (full ball, partial ball, or disc) is rotated in the flow stream to modify the amount of fluid passing through the valve.

Scotch yoke Device that translates linear motion to rotary motion. The linear motion is defined by a pin or roller element trapped within a channel in the rotary element.

Seat That portion of the seat ring or valve body which a valve closure member contacts for closure.

Seating torque Value depicting the turning force required to set a valve into its closed position.

Series Method of connecting the elements of an electric circuit so that the same current passes through them.

Shaft Portion of a rotary-shaft control valve assembly corresponding to the valve stem of a globe valve. Rotation of the shaft positions the disc or ball in the flow stream and thereby controls the amount of fluid which can pass through the valve.

Short-circuit current Prospective current which will flow under the conditions of a terminal short circuit.

Shutoff See *leakage*.

Solenoid valve Electrically powered device with a movable element that switches flow as a method of controlling actuator position.

Spring compression force Characteristic of a spring expressed as the restoring force available when the linear spring dimension is at minimum length.

Spring end torque Published value depicting the potential turning force of a spring-return actuator at the end of travel, when air pressure is zero.

Spring-return actuator Device that contains a spring element having the capability of returning a valve-actuator system to its normal position in the absence of air pressure, electric power, etc.

Spring-start torque Published value depicting the initial turning force of a spring-return actuator when air pressure is zero.

Starter Switch and/or resistance unit for running motors up to speed either steplessly or in steps.

TEFZEL Thermoplastic [25 percent ethylene, 75 percent tetrafluoroethylene (TFE)]. Used as lining and seat material in valves.

Three-way pilot valve Pilot valve used with single-acting (spring-return) actuators.

Throttling Action of a control valve in motion as it modulates flow.

Torque Product of a force acting upon a moment arm to produce rotary motion.

Transducer Device that senses a process parameter, such as temperature and pressure, and converts this information to an electric or pneumatic signal for use in control systems.

Transformer Basic transformer consists of two inductors (termed primary and secondary windings) wound onto a ferromagnetic core. If an alternating current or voltage is passed through the primary winding, a corresponding output will appear on the secondary windings in a direct relationship to the windings ratios.

Trim Internal parts of a valve which are in flowing contact with the controlled fluid.

Turning force Characteristic of a rotary valve resisting its change of position, or the output of a rotary actuator. Also known as torque.

UHMWPE Ultrahigh-molecular-weight polyethylene. Seat material offers high abrasion resistance superior to TFE. Temperature limit is 200°F (93°C).

Unseating torque See *breakaway torque.*

Vapor pressure Pressure at which a given liquid begins to vaporize.

Variable speed Characteristic of the actuator providing for revised (slower) operating time over the range of operation.

Vena contracta Location where cross-sectional area of the flow stream is at its minimum. (The vena contracta normally occurs just downstream of the actual physical restriction in a control valve.)

Volt Unit of potential difference or electromotive force. The final velocity of an electron which starts from rest and traverses a space across which a potential difference of 1 V exists is 593 km/s.

Voltage Potential force available to move electrons (current). When you start equipment which uses a great deal of electricity, the voltage may drop. However, if your power service and wiring are properly sized, the voltage drop will not be enough to affect the motor.

Wafer-style valve body Flangeless type of butterfly or gate, short face-to-face valve body. Also called a flangeless valve body; it is clamped between pipeline flanges.

Index

ABOUT THE AUTHOR

Wayne Ulanski is a vice president of El-o-matic USA Inc., a leading manufacturer of automated valve systems. He has been involved in new product development, market research, training programs, and managing design teams in the development of rotary-electric actuators.

A frequent lecturer and author on valve automation, his articles have appeared in publications such as *Chemical Engineering*, *Plant Engineering*, and *Measurement & Control*. Mr. Ulanski is also a course director at The Center for Professional Advancement, the world's largest private non-degree educational institution.

He is active in several professional and trade organizations, including Manufacturers Standardization Society, Valve Manufacturers Association, and the Instrument Society of America.